THE END OF THE MIDDLE

How Bots, People Analytics, and Remote Work will spell the end of management as we know it

Edition 1.0

30 September 2021

Peter C. O'Hanlon

ISBN 978-0-6450688-1-8

www.zerophase.ai/TheEndofTheMiddle

Acknowledgments

At the time of publishing, the novel Coronavirus and the global pandemic of COVID-19 that it has caused has already changed the complexion of our world and brought suffering to millions of people. For those of you reading that have lost loved ones as a result, I am sorry. This book deals with some of the business by-products of this pandemic. In no way do I wish to trivialise the devastation by focusing on a subject that could be seen as relatively trite - the nature of the workplace.

To the many managers that I have had the pleasure of observing and working with, you have shown me practices both good and bad. They have all shaped my thinking, and so I owe you.

To those of you that have plied your craft in the software engineering industry over many years, but have sometimes been ignored as unimportant 'code-monkeys'; you have pioneered many of our approaches to the future of work. Congratulations. I hope you feel vindicated.

To my family. Dad, thanks for being excited by this little project of mine. May you rest in peace. Dominic, you told me one afternoon that I should write a book. Who am I to argue? Finbar, your creative energy is inspirational.

To my dear wife Jen - thanks for your patience.

Who this book is for

The Leader

We are all familiar with the call for leaders to "walk the floor." It shows presence, lifts morale and gives the troops a chance to raise issues and concerns. It also gives leaders a pulse for what's happening at the coal-face, or at the very least an illusion of it. After all, nothing beats face-to-face human contact and a bit of unstructured banter. Or so we are led to believe.

The workplace is full of leadership norms, and management tips and hacks. Most are accepted as self-evident and lack any real empirical basis. Many have ceased to, or perhaps never did add any value. Will they be consigned to the waste-paper-basket, or undergo renewal in the days ahead?

Over the coming years, the mainstream adoption of AI and other data-driven techniques in the workplace will throw many of the existing practices out the window, and your role is set to change. As a leader, you will have the opportunity to be more directly connected with the front-line at work, and will be expected to do so. Despite your busy-ness, AI will help you do this.

If you are a leader that wants to know how, then this book is for you.

The Employee

As an employee you are probably already aware that AI-powered automation is replacing blue-collar jobs, but the effect extends beyond the industrial and manufacturing sectors, and is starting to impact white-collar professional roles in the knowledge and service sectors.

This book affirms that this change is accelerating, but also recognises that many jobs and their requirements will change.

How you contribute at work, and how you are managed will be fundamentally altered.

If you are an employee that is curious about the future arrangements of your working day, the means by which it will be directed, how you will relate with your colleagues, and whether you will be reporting to a robo-boss, then this book will give you a vision of that future.

The Middle-Manager

The Middle-Manager has a complicated life, bridging the executive with its strategy, and the front-line and its operational execution. They broker the compromises, being sufficiently across everything - or at least they are meant to do so. In a world dominated by bots, AI and autonomous teams, will your role become redundant, and if so what path should you pursue? Should you seek to become an executive as quickly as you can, re-skill in a delivery role that is insulated from AI, ride it out, or focus on the unique things that you do that no machine will ever successfully replace?

If you haven't answered these questions yet, then this book will help you do that.

The Human Resources practitioner

In the days ahead, advances in AI will mean that many roles will be deleted from the org chart, and bots will become prominent. Managing this inevitability will be a challenge. For the workers that remain, their relationships with bots will become strangely complicated. In society, people have started building fun, entertaining, productive and romantic relationships with machines, and it will be no different in the workplace. Your specialised training in human resource management will help you play a vital role helping navigate this journey. This book will give you an in-depth understanding of what these changes will be, and a view of what opportunities and risks that the emergence of bot-workers pose, as you build the future of work

The curious

This book is for anyone interested in fresh perspectives on the People Analytics, AI, bots, agile methods, digital transformation, the future of work, and employee experience. It explores these topics, giving an overview of each, but at sufficient depth to be a little bit dangerous. It also ties them together to show how a data-driven workplace will birth a new business architecture that I call the Conversational Enterprise.

I hope it provokes some fresh thinking for you.

INTRODUCTION

Multiple factors have been conspiring to change ***how*** office work happens. The abrupt shift to working remotely has accelerated the adoption of new collaboration technologies and flattened hierarchies. AI can learn patterns that help machines mimic human behavior so quickly that swathes of jobs have been lost to automatons. The strength of relationships between workers and their employers has diminished as business has become 'Uber-ized' to the point where organizational boundaries are hard to see, and subject to legal battles. The confluence of these issues point to a fundamental change ahead.

The bulk of research focussing on the future-of-work (FOW) relates to how these changes will impact the individual worker, and which roles will survive automation. In this book I focus more on the impact of algorithms on the management of organisations, and the future place of managers in particular.

Some of the ideas that I cover in this book, and the conclusions I've arrived at will be controversial. A portion of you will celebrate, and others will likely take offence. I am not seeking either reaction. Instead, my goal is to project into a not-so-distant-future to identify the threats and opportunities, so that you and the community more broadly can prepare for what's coming.

If as a result of reading this book you feel more at risk, I believe that is a good result. Knowing what's ahead may help you avoid a dismal outcome. If I have painted a future that you don't care for, I think that too might be useful. There are elements I don't like either and I hope to be proven wrong. Please join with me in opposing them. If you are excited or motivated by what you read, I am also delighted. There are some fantastic opportunities ahead.

In writing this book I have tried to steer clear of science-fiction based, long-term, hypothetical scenarios and have focussed instead on change that is already underway, with technology that is currently at work, and that we may expect to see fully embedded within a decade.

In terms of the flow of this book, I have attempted to make chapters reasonably self-contained so that you can dip in and out at any point. However, they are ordered with a particular argument in mind. Initially, I recap some of the major changes currently disrupting work, and explore in this context the place of mid-level managers. Next, I detail the purpose and capabilities of three data-driven innovations; conversational AI-bots, People Analytics, and Agile HR. Having established what they offer, I propose a new approach to management that draws on each of their strengths, and make some predictions about what that will mean for management (*hint*: the need for managers will diminish). I conclude in practical terms, exploring the nuts-and-bolts of how specific work practices are likely to change as a result.

CONTENTS

WORK HAS CHANGED FOREVER

Major disruptions to the nature of work are happening on a global scale. Digital workplaces, the rise of AI, and remote work have coalesced to redefine the working day. At the intersection of these changes, we see clues to the fact that not only are some jobs ripe for automation, but that we are on the verge of more fundamental shift in how work is organised.

The last gasp of industrial age management

The Fourth Industrial Revolution is changing how we work, but our management practices have been imported from previous eras.

My father was many things, to me and to others. In his earlier years, one of those things was the superintendent of a cable factory. Within its walls, workers were organised into shifts; raw materials would find their way onto the dock and be moved into the production line. Other workers would be changing settings on machines, inspecting quality, and managing inventory. Items would be shuffled from one process to another for further transformation.

Of course, the whole operation required a large roof and several walls in which people and machines could be coordinated, in order to keep the lorries (as they are called in the UK) delivering on time. At homes throughout the countryside, alarm clocks would spring into action so that workers had time to get their uniforms on, fuel themselves for the day, and get to their expectant machines. With disciplined routine, they would arrive in time to clock-on, go to their stations, and start their duties. My father looked after a few hundred such workers. He would talk to them, review their work, and give direction. Sometimes his superiors needed him to relay new instructions. That was his job. He took care of his workers' needs and did everything he could to ensure that the factory was meeting its targets.

This story is not unique. It has been repeated in one form or another for several centuries, ever since the start of the industrial revolution. Much the same model of management has been applied to cotton mills, coal mines, sprocket factories, and increasingly to a range of service businesses, such as banks, insurance companies, and civil service departments. Throughout this 'four-walls-and-roof' period, it was easy for managers to eyeball whether their teams were meeting production and quality targets, and to intervene to ensure that they were. If there was a problem in one part of the human-machine system, it was a simple matter of observation and adjustment. Everything was physically in one spot.

We live amidst the fourth industrial revolution (4IR) which is introducing change at increasing rate. Businesses operate 24x7 in multiple locations, connected by computer networks moving electrons at light speed. People and machines are integrating their powers to generate new consumer experiences. Leaders across industry verticals including banking, manufacturing, publishing and retail have adopted a view that they are now, a tech company first of all. Robotic automation takes care of manufacturing and fulfilment while human overlords issue instructions from whichever town their stand-desk happens to be in. Whatever the business, populations consume their services directly from their lounge-room chairs and through their screens, interconnected as they are. Once distinct businesses are all blurring into platforms. Who would have thought that a bank would conceive of itself as a technology business (PYMNTS, 2020)? Who would have thought that the world's biggest' hotel chain' was a software development company (McDermid, 2017)? Who would have guessed that the biggest advertising revenues would go to a text indexing platform (Kollewe, 2017)?

Alarm clocks don't ring in unison in the village anymore. There are no queues at the gates for people to join their mechanical workmates, and people don't bark instructions from vantage points from which they can observe. No longer just a hallmark of early high-tech digital businesses, office workers now blur their backgrounds on video calls, and tap away on keyboards to chat as network lights twinkle away indicating the transmission of data. The 'four-walls-and- roof' model has given way to ethernet, wifi, and a 'walls-optional' mode of working.

While many aspects of work are no longer recognisable, our management practices resist change, and remain mostly the same as those that ruled the factories and cubicle-farms of the past. It is time for management practices to catch up.

- **How effective are the management practices in your workplace, and what actual evidence do you have to support your view?**
- **Are your management models imported from another era?**
- **What would happen if all your managers stopped coming to work?**

Working from home has sparked changes to work practices

The shift to working from home has accelerated new thinking about how to organise work, and the adoption of technologies necessary to support it. But working in a distributed way is not new. Analysis of how well-established industries have supported their workers reveals the critical ingredients for survival.

No change to work accompanying the COVID-19 global pandemic has been more obvious than working from home. The initial lock-down periods saw businesses respond quickly, becoming more operationally functional with staff working from home than they thought would have been possible. All the predictions are that there will be no going back (e.g. Senz). In a matter of weeks, many businesses had equipped their staff with the necessary tools to get the job done in isolation. Tools that enabled staff to be together digitally like Zoom, Microsoft Teams, Mattermost, Slack, Twist, GitLab, WebEx Teams, Atlassian etc. experienced strong growth. People reported "Zoom Fatigue" as a result of being on camera for most of the day, and people either dressed up their home work spaces to give a professional glow, or switched the background blur on. Business folk unfamiliar with chat applications learned about teams, channels, streams, notification and became adept at using social collaboration tools instead of (in reality as well as) email. Companies already familiar with distributed work technologies experienced an increase in productivity ("The State of the Octoverse,

2020", Github). What's more, with staff equipped with new technologies, many companies have since reported an increased appetite for further and deeper business change (Baig *et al.*).

Workers learnt how to communicate quickly, but "remote" work has also meant that businesses have had to come to grips with how to coordinate work, with many of the processes that they used to rely on for a colocated workforce no longer being possible. For example, with people working at different times and locations, it's hard for all decisions to be made and communicated from the centre. To avoid introducing delays, the flow of work relies on coordination rather than control, and that individuals and teams operate with increased freedom rather than being directed in a top-down manner. There is a need for some give and take between the units of delivery and the enterprise as a whole so that autonomy is balanced with alignment. New processes and technologies for coordinating this interplay allow for work to be designed, completed and tracked with 'radical' transparency (Gitlab), providing a more social approach to achieving organisational outcomes in a distributed manner (Ingham). Teams are now equipped to contribute to business outcomes, rather than teams having to be 'managed' by personnel dedicated to coordinating them in that mission. Some businesses are yet to fully adopt a distributed model for remote work, but most are at least a little way down that road.

The relative merits of remote working, and its implications on employee experience have been the subject of research efforts for several years. Many businesses have now run their own surveys to find out what their staff think about it. Unsurprisingly the results are often mixed with positives such as reduced travel time accommodating better work/life balance and negatives such as feelings of isolation. Despite this growing knowledge, the impacts of remote work on management are not as well understood (Hinds and Kiesler). One place to start in discovering what works is by performing a scan of those industries that have had a 'remote' workforce for decades. Doing so reveals some common and necessary survival patterns.

Taxi!

For taxi operators, the tasks of daily work are short-term, and hopefully plentiful. Taxis receive job notifications from a suite of information brokers that serve up the opportunities. Drivers follow the work at the times that suit them best. Their cabs are full of displays, screens, and more mobile phones than the neighbourhood drug dealer (which we'll come to later). They have flexibility with regard to the times that they work, but little else. The networks link them all together to achieve a system that can support employment, get people where they need to go, and make a profit. The taxi industry is a classic remote workforce that maintains constant real-time contact over proprietary (as well as personal) channels to make it all come together. Without access to information, operators would be relying on experience, guesswork, and hope for their next task. Access to the data (rather than a management layer), with a strong incentive structure allows individual operators to use their judgment and effort to compete with one another. Apps such as Uber, DiDi, Lyft and the like have further disrupted the entire transport sector by connecting riders and drivers more directly, integrating riders into the network (*note* - no need for humans in the middle).

Crime gangs

Achieving business resilience over long supply chains, under significant competitive pressure, and ongoing regulatory opposition is hard. It's such a big challenge in fact that it attracts great talent, willing to take risks chasing large financial rewards.

Criminal syndicates have been refining their business model for hundreds and maybe thousands of years, and usually operate in a distributed manner. With carefully designed organisational structures, mobile phone "burners" and encrypted messaging platforms, these networks control supply chains with sophisticated financial and technological support. They rely on strong command and control structures based on powerful incentive (and disincentive) schemes. You've seen the movies. Their structures intentionally preserve its members' anonymity so that if individual nodes in the network are taken down, damage is limited and the syndicate can carry on. The dark web strengthens the connections between the players, and further removes reliance on brokers. Limiting corporate knowledge in this way further protects the enterprise from its competition (including but not limited to the law).

Organisational design, information flow, and setting up suitable incentive structures are core competencies for criminal syndicates, and innovation with different communication channels provides a competitive advantage. Inter-personal control needs to be exerted at a local level, and its necessary for the fear of a bad 'performance review' to cast a long shadow on future behavior. This allows face-to-face pressure to be exerted relatively infrequently.

Freelancers

Freelancers are specialists with relatively rare or hard to acquire skills, usually in activities that are only occasionally needed by individual businesses, but frequently enough across an industry or market. The freelancer aims to pick up enough gigs to flatten the cycle between feast and famine. Many of them use digital channels to directly access clients across a global market. Their apps help them bid competitively, clarify client needs, supply incremental deliverables, and deliver the final output. The freelancer has serious flexibility in when, where, and how they work. Beating the competition means winning the war for ratings, and shaving costs, which can of course lead to lower standards. Like the taxi industry, their manager is an app. Some industries (such

year, and so moving to remote work never needed announcing. Instead they have a very particular and unusual culture and mode of operation that was designed to sustain working that way from its outset, which has been reflected in their software platforms.

Other tech businesses have experienced the highs and lows of remote work. Marissa Meyer, a former CEO of Yahoo, famously decided to dial back remote work in 2013 in favour of core times and use of offices (Arthur). Satya Nadella, CEO of Microsoft, has warned of the risks of a permanent move to working from home leading to social isolation and negative mental health consequences (Manfredi). While some tech businesses prosper without people ever being physically together, others have struggled with it; and those CEO's now seeking to bring their staff back together are at their "wit's end" (Belvedere). They have all had the necessary tech run place, but their success with this mode of working has depended on their work practices, strategies, and cultures.

The DAO

An extreme example of a distributed organisation was The Decentralised Autonomous Organisation (DAO). Established as a vehicle for investors to back projects, it had no central authority, no management, and was purely a set of contracts. People coordinated their efforts in a transparent and auditable way with the goal of funding investible projects without the perceived 'interference' and misdirection of funds from fund managers. Investment decisions were made entirely on the basis of voting rights of the owners of funds. However, the technology that it was built on, and the value of funds that it was directing made it a target for attack. The lack of a management structure made the response difficult to coordinate, and regulators were uncomfortable with a stateless, director-less organisation. It was a bumpy six month ride for all involved, and is a fascinating example of how advances in technology can pique people's strong desires to

as food delivery) have been disrupted, with employers implementing a digital management layer. They have argued that workers are freelancers rather than employees, but regulators are starting to disagree (Deutsch and Sterling, 2021).

Succeeding in these industries over the long haul means that tasks need to be well understood, tracked, and delivered efficiently and repeatably: all mediated, controlled and incentivised by the data produced and surfaced in the tech.

NASDAQ and friends

In May 2020, Twitter announced that its workforce would never need to come back to the office. Have you wondered what it would take to be able to make such a call? At the time Twitter's revenue was $3.5bn, had 321 million active users, and had 4,600 employees centred around 25 offices worldwide. It is hard to describe the 13 and a half years growth since its founding as anything but astronomical, and the organisation as anything but large and complex, and yet with 280 characters, Jack Dorsey could let everyone know that where they worked (physically) was irrelevant. It is no surprise that as Twitter was a business based on creating tools for connecting people that its workforce would be well placed to do so remotely. The change was not as sudden as it may have appeared, and not as radical as it was for many other businesses. Since day one, Twitter had been building a digital business, where innovation could happen remotely, using tools and processes that allowed employees to connect across the world in different time zones in standard ways that enabled efficient work. It was not optional for them - their round-the-clock operations mandated it. What may have looked like a dramatic shift outside was met with a shrug from most staff.

GitLab (global leaders in distributed work) intentionally structured their operations this way since starting out and have never had an office presence. Yet they continue to grow year-on-work together, but not necessarily under the control of a central authority. There was no unifying incentive or means for the parties to work together in helping the organisation survive a crisis.

- **Do your people require real-time connection with one another, and do they have it?**
- **Can you see parallels from other businesses types that have been operating in remote and distributed ways for years?**
- **Are there elements of your business that will never change because of remote work? What makes them so insulated from it?**
- **What is the optimal balance between remote, colocated or hybrid models. How would you know if it needed changing?**

The rise of AI and automation

Artificial Intelligence has finally broken out of being a niche hype-based capability, and become mainstream. It is developing at eye-watering speed, but how it is applied in the office is yet to be fully realised. We don't understand how it will impact management, and to this stage, the subject has been mostly neglected.

What is it about Artificial Intelligence that caused the likes of Elon Musk and Stephen Hawking (hardly luddites) to contribute to the "Open Letter on Artificial Intelligence". In it, they alerted the globe to their personal fears of the potential demise of the human race under the "hand" of AI.

Intelligence is typically held to be a positive attribute, enabling species to compete, with humanity currently sitting at the top of the chain. But what happens when networked machines, with their vast compute power use algorithms at the speed of light? The fear is that our biological circuitry, which has evolved over hundreds of thousands of years simply wont be able to keep up. We may inadvertently be creating a technology that becomes a runaway overlord, a scenario in which we might end up as the new pets, playthings or just simply surplus to their silicon needs. This sounds like a movie scenario, but our best and brightest argue that this is a real potential and not so far over the horizon.

You might dismiss these claims as alarmist, favouring instead the view that AI is just one of those relatively benign applications on your smart-phone. This is understandable for a couple of main reasons.

1. Many of the systems labelled as AI are no more intelligent than they are supernaturally magic, which is to say not at all. Motivated by commercial interests, businesses tout deterministic if->then->else rules based systems as “AI”. These are not the kind of systems that Musk, Hawking and their contemporaries are concerned about, and they won’t out-smart us any time soon.
2. AI isn’t new. Many of the major theoretical underpinnings of AI were developed through the 1950’s and 1960’s, and people have been highlighting the potential risks of AI and computers becoming “self-aware” since this time. AI has been hyped as “coming” for a long time and we are all accustomed to hearing the threat, which seems perpetually on the horizon. However, it wasn’t until massively powered computing was commoditised in cloud powered the explosion in the scale of data storage and processing that AI has emerged, and revealed itself as a practical reality. GPT-3 (a n AI-backed language model built using 1.5 billion data attributes to autonomously generate sensible text) was simply impossible to create with punch-cards.

Progress in the last decade has given rise to a suite of applications of AI that we have become so accustomed to (including navigation, speech recognition, fingerprinting, and video filters that make people look like cats on conference calls) that we barely identify them as such. We are now at the stage where technology has become invisible, that living without it is almost unimaginable. As we carry yesterday’s supercomputers in our pockets, with a full array of sensors, AI systems are being deployed everywhere, and are finding their way into most of the corners of our life.

Facial recognition, once the exclusive capability of security agencies and TV sci-fi is now such a commodity that it unlocks our phones. Apps that perform automatic language translation live means that Douglas Adams' babel-fish is a reality. As we buy online, machines consider transactional, cyber and geographic details to highlight and park anomalous transactions, making widespread online retail financially viable. The books that were filled with street maps in the glove compartments of cars are gone, as our digital assistants tell us how to get where we need to go, re-routing as we go because of traffic snarls we can't see yet. AI is now ubiquitous and hardly noticed.

AI can beat humans at increasingly complex games ("AlphaGo: The story so far."), can out-design and will replace 90% of architects ("Rise of Artificial Intelligence Means Architects Are 'Doomed' Says Sebastian Errazuriz."), and can outperform humans in Stanford's reading comprehension tests (Fenner). It's starting to look like a one-sided performance of Annie Get Your Gun's "Anything You Can Do (I Can Do Better)". Operating at scale, its pure brute-force allows AI to develop insight that makes it seemingly unbeatable.

As we shift our lives into digital platforms, more data is created, supplying increasing volumes of raw materials for our AI systems to learn from. Mark Cuban, outspoken tech investor has described businesses as either AI-haves or AI-have-nots, principally because they will compete on the basis of their AI capabilities. The data we generate that will build tomorrow's AI is considered so valuable, that businesses provide their services to consumers for free in order to gain access. Business sees that AI will shape the future and has banked on it.

And in case it wasn't clear yet that the genie isn't going back into the bottle, training in AI is now one of the most in-demand educational courses that institutions deliver. (*side note*: in my opinion because the process of training AI systems is highly repetitive and automate-able, machines will be training machines in no time). AI is ubiquitous, powerful, and the data sets they consume, and skills required to build them are increasing.

In face of this expansion, and the risk that tech might outrun our capacity to control it, groups such as OpenAI have emerged. Their focus is on progressing the frontiers of AI in a safe manner, by providing education, along with carefully controlled access to leading-edge capabilities. How this will all play out is unclear, but the game is currently in progress.

As a society, we are yet to see how such AI systems will transform work. The applications of AI in the office have largely been limited to automating workflow processes, and removing manual repetitive tasks such as data entry. For example, up to 47% of jobs are considered to be at high risk of automation by the mid-2030s, and mostly those occupied by workers with lower wages and educational attainment (Frey and Osborne), but the implications are more far-reaching than redundancies. machines can now out-think and out-create people. The ubiquitous chatbot is shifting from being a dumb 'if->then->else' processor to an interpreting, researching, thinking, synthesising, machine. In remote workplaces where people interact with one-another as avatars in a collaboration systems, there is potential breakdown between the distinction between humans and such machines, with far reaching implications.

- **Do you see applications AI in your personal life more so than in your work/professional life? Why is that?**
- **Is AI a fad? Do people treat it as such in your business?**
- **Do you have any views about how AI will change work in your organisation? What about your own work?**

MANAGERS ARE VULNERABLE

Most analyses of the threats posed to jobs from AI-driven automation focus on blur-collar work, and office jobs that require lower levels of education, whose tasks are relatively repetitive in nature. There is an expectation that professionals and managers will be more insulated from the change. However, in the office managers are amongst the most vulnerable roles. While the impact may be resisted due to their position of power, tenure, and chance to protect themselves, the transformation is both sizeable and inevitable.

Under threat

For many years, businesses have been thinking of ways of exercising a necessary degree of control over their workforces without relying on managers, motivated by the belief that managers are ineffective, expensive, innovation stifling, and organisationally stratifying.

There was a time when the only way to exert control over an organisation was to hire managers. As businesses grew in size, it became untenable for the executive to be everywhere. Managers were needed as the 'span of control' became too large. As enterprises got larger and more complex, more managers were needed. As the number of offices multiplied across locations, more managers were needed. As specialist domains like finance, human resource management, risk management, and others emerged, more managers were needed. It has been a simple formula to support growth. As interesting parallel, research shows that as economies develop, the percentage of the workforce that occupy manager roles also increases (Fix). The proliferation of managers seems inevitable.

Because management has been a growth area, being a manager has been a good career path, and many workers have spent their careers in the ranks of management. Having started out in a front-line delivery role, their competence in their specialty coupled with a natural ability to communicate has seen many promoted into management, and live out their days doing management-y things, usually on a higher salary. The path to being selected as a manager is often a function of tenure, or simply that fact that someone deserves 'a turn'. These facts at least partly explain why 82% of manager selection decisions are poor ones (Beck and Harter).

But in recent years, the situation has been changing. Traditional management is at breaking point and managers are burning out. 81% of managers say their job is harder than it used to be and 63% want to get out of management (Beauchene and Cunningham). It seems that the rapid rise in 'complicatedness' at work, with the number of 'fixes' or rules added to overcome workplace problems and unusual corner-cases is outstripping the number of actual challenges that they were mean to solve in the first place (Morieux). The appeal of being a manager has declined for incumbents, and being the manager is starting to be equated with "drawing the short straw" amongst those in the next generation with only 9% of non-managers want to be a manager (Beauchene and Cunningham).

Management has been described as "the least efficient part of business" (Hamel). The effect of managers is often to create unnecessary work, slow decision making, disengage workers, stifle thinking, resist change (e.g. Juneja), and confuse objectives. Approaches to 'empowering' workers to operate with more independence is an accelerating trend. And because this complicatedness has prevented people from doing work, there have been numerous attempts to put a lid on the swelling management class, and devolving 'management' activities throughout the workforce.

- In the 1950's Theory X and Y was advocated as a more open approach to management, by focusing on building motivation, an intrinsic incentive to work well. This management style encouraged people to dig into their own reserves of discretionary energy, and to connect their own work with the mission of the organisation. Many of the characteristics of this management are still seen in contemporary leadership.

- Based on the positive military experience of loosely coupled networks of highly cohesive teams, McChrystal popularised the concept of a 'Team-of-Teams' in business instead of a management hierarchy of command and control. If it can work in a military, where rank is so important, it can probably also work in business.
- Approaches to software development have led to the emergence of Dev(elopment)Op(eration)s teams. DevOps teams contain all the skills required to deliver a service, operating as a tight knit and autonomous group. Alignment between these teams is achieved through networks of people and processes that integrate their efforts as a system rather than by a series of managers with cascading ad hierarchical responsibilities. This approach was described early on by Spotify's famous 'Engineering Culture' documentation, and its set of non-linear career 'Steps' that eschewed competent specialists having to find a management role in order to fund the needs of their growing families.
- Hamel (2011) describes the libertarian approach at The Morning Star Company, the global market leader in tomato processing, where there is no central planning authority. Despite the size of the operation, budget decisions are made in an open democratic manner through discourse and voting. People write their own job descriptions and compensation is based on peer reviews.
- Economist Yannis Varoufakis (Oct, 2020) extends this kind of thinking to lay out a vision for an economy of extreme democracy and a lack of concentrated power where ownership and labour go together. Whether you agree or not, he cites amongst his influential experiences a stint at a

software business that worked in a similarly democratic way.

- A formal attempt at codifying a 'no structure' organisation is holacracy. In it authority is decentralised and dispersed throughout an enterprise by focussing on process and role clarity. It promotes a system of organisation that relies on a set of core principles tailored specifically to each organisation. Whilst Zappos has been a high profile adopter and advocate of this system, it has numerous critics that claim it leads to chaos. I'd like to see it in action as a reality TV show.

Managers are under attack, and the range of attempts to create businesses that don't require them is a symptom of this. If a bigger sign of this issue is needed, the emergence of the gig economy, and the rise of businesses such as Uber (that try define themselves purely as a marketplace for freelance contractors) should suffice. In the media, the focus of criticisms of such businesses has been that they are a crude attempt to minimise their legal obligations to workers, leaving the lowest paid to fend for themselves. This structure also has the usually overlooked impact of obliterating the management layer that would have ordinarily existed. These huge cost reductions make them highly competitive. For many now, the drivers' boss is an app that exerts control through tasking, and performance measurement (from passenger ratings). There is no need for managers in such an enterprise.

One common element in attempts to curtail management is the implementation of processes and communication patterns that make work visible, trackable, traceable and sometimes rateable. While Morning Star are lo-fi and physically colocated, most office-based businesses deploy a technology backbone to facilitate autonomy whilst embedding alignment. People can work more effectively across teams without hand-holding and channels exist to support cross team discussions. Decision making that includes expertise can happen without having to go up and down the chain, across the silos, and drafting papers and running meetings. In a workplace like this, presenteeism iOS not rewarded. There is little opportunity to fake your value by always being there because work outputs are transparent - a point we'll pick up on later.

While there is a pent-up demand for manager-free or low-manager organisations, the vast majority of businesses still rely on a centralised hierarchy to organise itself. And not all efforts to reduce reliance on managers have been successful. As an example, at the request of engineers, Google conducted an experiment to do away with managers. It didn't go well, and conflict arose. As is often the case in high-ego competitive environments, people tend to overestimate their importance, and underestimate that of others. Simply removing managers without supplementing their positive influence is a recipe for failure.

Quality middle management has been shown to play a number of vital roles in the workplace. They can influence motivation by providing people with a personal 'why' to give their best at work and can be key in managing conflict and fostering productive relationships, as well as modelling positive behavior. Advocates of the value of middle managers argue that managers' psychosocial human qualities make them indispensable (Greenwood and Krol), and that their main contribution goes at work goes well beyond coordination efforts. While the case can be made that managers are indispensable, the economic incentives to keep searching for ways to replace remains.

- **Do the managers in your organisation get the best out of your organisations investment in talent? Who is expected to do that?**
- **Do managers have too many other duties to effectively manage their people, and are they the right people to do this?**

- **How are middle managers coping with work since the COVID pandemic?**
- **What effort or investment goes into supporting management?**
- **Are managers leading change or resisting it, and what is being done about that?**

The mathematics of replacement

Automation and replacement of front-line roles will have a cascading and disproportionately large impact on the middle-management layer, causing it to shrink by an even larger factor.

This chapter contains a little mathematics because the numbers speak for themselves. Consider the following.

Business A is small and has a team of 10 workers. For simplicity lets say there are 11 people all up, and one of them manages the other 10. HR folk call this number the "span of control". This is a reasonably high number in the office context, but it makes the maths easy. The ratio of managers relative to team members in Business A is 1:10.

Business B is bigger and has 10 teams of 10 people each. That means it also has 10 managers, but those 10 managers need a manager too and so, Business B has 11 managers. Now the ratio of managers compared to team members is 1.1:10, and there is an extra layer of management. As the number of team members grow, the number of managers and layers grows just a little bit more each time as a percentage. The increase doesn't look like a serious issue at first, but it compounds, and when measured in cost terms, it's substantial. Not only does the percentage of managers grow in relative terms as a business grows, so too do the layers. And the cost of the individuals in each layer increases as well. Again to keep the maths simple, if the cost of an employee doubles for each layer that they sit up in the hierarchy, the relative costs of managers for Business A are 20% and 24% for Business B.

But in an office, you'll often find a span of 5 reports per manager (including those managers in specialised domains with few staff). Using the above calculus, once a business reaches about 5000 people, 50% of its labour costs will be managers. That's a big juicy target.

Further compounding this cost, as the ranks of management swells, so does the cost of completing units of work. With silos come delays. More layers means more 'papers', and the Tetris-like task of arranging meetings means they happen infrequently, reducing nimbleness and increasing opportunity costs. Let's assume that those at the top of an organisation meet once per month to make decisions, and each layer below does the same. It doesn't take a lot of imagination to see how the approval of an idea from a team member can take months to be approved as it is filtered, improved, massaged, socialised, packaged, forgotten, and rediscovered before approval. Rather than streamlining work, adding managers often introduces new work. In a dynamic described as 'Organisational Drag' (Mankins and Garton), managers tend to create work for others to do, supporting the organisation of their schedule, drafting and editing papers, adding regular meetings and introducing change. The effect can be a perverse, self-perpetuating bureaucratic one, with managers creating so much work that more staff (who then also require management) are required to get those things done. Management bloat needs to be controlled if costs are.

The inertia and cost brought about by having a large cadre of managers is huge. Large organisations, with complex functions usually require an army of mid-level managers. This imposes a tax that limits investment in things. So there is an unavoidable business case and focus on reducing the negative impacts of middle management. Perhaps this is why 37% of managers believe that this layer will disappear within 5 years (Beauchene and Cunningham).

How does this look in a shrinking workplace where front-line roles are cut? We can easily see by running these equations in reverse. As cuts are made, those businesses that have established a "house-of-cards" management system will be exposed. Just as the management class has grown disproportionately, if there are fewer people to manage, then the management class will also shrink disproportionately.

- **Have you considered the cost of management activities in your organisation, and whether they add value?**
- **Do you believe that management activities in your business are efficient or inefficient?**
- **What is the ratio of managers to workers? Is that appropriate for your organisation, and how do you know?**
- **Does your business have management bloat?**
- **Are decisions made with nimbleness, and if not is that evidence of bloat?**

Managers are struggling to do the 'human' things anyway

There is a view that managers are safe from automation because they have special and necessary human skills that AI will never be able to replicate. In reality, management is struggling to do these things anyway.

Many point out that managers play a critical role in organisations because of their uniquely human qualities (Greenwood and Krol), and by implications, machines will never replace them. Time will tell whether or not this is the case.

Irrespective of whether bots will be able to accurately emulate human qualities or not, managers are struggling to demonstrate the human qualities that are meant to make them special anyway. This deficiency has been accentuated by remote work, and so the bar for replacement of managers by bots is in reality, surprisingly low. The distinctive qualities that humans bring to management in practice is often not that impressive after all.

The adage that 'people leave managers, not companies' highlights the importance that workers place on their relationship with their manager, and that managers often fail to give workers what they are looking for. Over the years, research has shown that managers have an impact on performance, morale and attrition, but a recent study (Gallup Inc 2015) has shown that the manager relationship is the single most significant factor of all, and accounts more than 70% of the variance between employee engagement scores amongst staff. In this work they show that people are often dissatisfied with their manager because of what might be described as distinctively human qualities including conflict, favouritism, apathy, incompetence and personality. These aspects are often the motivation for why people leave their manager and so perhaps the idea of having a robs-boss isn't as silly as it sounds.

In truth however, the adage works both ways, and employees can stay in sub-standard companies because of the excellent quality of their direct manager. On the bell-curve of manager quality, good managers retain people and bad managers shed them, with most being neutral. What is most concerning though is that this same research also found that 87% of employees world-wide are not-engaged or are disengaged (that is actively and knowingly not caring). The bell curve is not centred on "OK manager", it is centered on "Bad manager" with "Good manager" at the rare positive end of the curve. These statistics suggest that on average, humans are not so good at doing the 'human' things very well. It doesn't offer much succour to the argument that machines could never replace them due to their lack of humanity.

Of course manager quality varies from one to the next, and if you ask managers which part of the bell curve they are on, you don't even get close to an accurate answer. How managers perceive themselves differs wildly from how their workers see them. When asked in a recent survey (Bolden-Barret), 80% of managers claimed to be transparent with their teams, but only 55% of their employees agreed. When asked about whether managers personally cared about employee well-being, only 53% of employees agreed, and 80% of workers believe that they can effectively do their job without their managers.

Sadly, this situation is likely to have been made worse with widespread "work-from-home" happening as a result of the COVID-19 pandemic, with research revealing that a large number of managers have been finding remote management more difficult, with employees interpreting this as feeling increasingly untrusted and micromanaged (Parker et al).

Fault doesn't all lay at the feet of managers. In their defence, the same research shows that managers are so busy that they don't have the time to invest in individuals, and lack the skills to perform their function well. It is also claimed that fewer than 50% of managers have received any formal skills training (CIPD), and even though $356b is spent globally per year on manager training, it doesn't seem to work (Beer *et al.*). Coupled with the fact that 82% of managerial appointments are poor selections (Beck and Harter) it's hard to blame them.

These statistics show a poor state of management in general. Rather than supporting their teams, and effectively bridging between the executive and the front-line, they are getting in the way. There are multiple causes but whichever the case, human managers are struggling to bring their human selves to work and do the human things that are meant to make them indispensable.

Are managers under threat from AI? Most predictions are that management is a profession that is actually under low threat from AI. For example the website "*Will Robots Take My Job?*" rates this profession to be at what it considers a small risk (25%) of being negatively impacted by AI in the next two decades. In reality they are already under threat, and efforts to improve their performance through training and better selection haven't changed the situation.

I believe that predictions of job security are wrong, and that there is a bumpy ride ahead for those in the management cohort. Sadly, many may find themselves unemployed, not purely because of AI, but because it tipped the scale just that little bit further.

- **What do the worker in your business think of their management? Do they perceive it as highly functional, and worthy of trust?**
- **What kind of impact do you think the managers have on the workforce, and does this impact support your organisation's goals?**
- **Is your management exclusively focussed on the tasks and decisions of work? Do they get involved in improving how people do their work?**

Management will eat itself

The introduction of new technologies has enabled teams to self-manage effectively, and align their efforts with organisations strategy. This has reduced the scope of the role of managers. What remains of their role are the human aspects of work, providing motivation, context, guidance, listening to their teams, and bubbling up concerns. These aspects are also ripe for disruption.

In traditional businesses, middle management acts as a bridge between the executive and the front-line. Its function is to listen, interpret, communicate and influence up and down. It works as a link in the chain of command, to align the parts of the business in executing its strategy. They need expertise in the tasks that their teams performed, but are also *au fait* with the broader organisational context. This allows them to make daily decisions about sequencing work, and defining how things are to be done, balancing the resources needed.

Remote work triggered the massive adoption of collaboration tools as businesses scrambled to keep their businesses functioning while people stayed home. There are now literally thousands of apps to help keep teams humming in their daily tasks; direct messaging, team chat, voice, video, screen sharing, wikis, document management, and task management have all become standard (if not mandatory). These tools have enabled teams to

channels. The race is on to own the entire office worker user experience.

A unified work platform would allow the worker a lower distraction and relatively friction-free work experience, but given the diversity of systems and tools being integrated, a design approach that unites them all into a common experience is needed. Without it, employees would still experience the jarring effects of jumping between different back-end systems, with their various user interfaces, with the conversational user interface (UI) acting merely as a portal or launch point.

A major contender for this pattern is to bypass a visual interface altogether, and rely on a "conversational interface". Conversational UI works by reorienting the experience away from a predominately the navigational menus, buttons and links (the dominant approach to interface design), to a more natural and human one, reliant on human language (Schmidt). Rather than tabbing, clicking, and filling forms, users would simply talk to a conversational agent (a bot) that knows how to operate all the back-end systems. For example, instead of opening a specific project management application to set up tasks and update them, users simply tell a bot a project status and the bot enters all the updates bypassing the user interface. The bot may need to clarify what is needed, but it will go away and do what the user needs on their behalf. Users won't need to learn user interfaces. Instead of logging in to an HR system to apply for leave, users will just ask a bot to do so for them. Instead of clicking a link to approve requests, people will simply be asked by a bot if they approve something. In the background the bot will manage all the necessary records. The examples can go on *ad-infinitum*. Siri and Alexa work on a set of standard consumer tasks and the same thing is coming to the enterprise. My opinion is that some domains such as the trading-floor, the cockpit, and the surgical theatre will still be dominated by densely filled screens with visuals and data rather than chat conversations (the human brain can process a lot of information really fast when presented visually), but the tech community seems to be betting on a future in which office workers interact with all their applications with human language (e.g. via bots).

Conversational UI is a reactive use of bot technology, responding to users needs. As we shall see, bots as able to initiate activities of their own, and turn the tables to achieve other goals.

- **How many software applications do your employees have to use to get their jobs done?**
- **Is your business adopting systems that work together cohesively, or are they still disconnected?**
- **Is the data generated by your staff as they work accessible from your current analysis platforms?**
- **Is anyone looking after the full digital experience for employees?**
- **Have you considered how conversational bots might offer an alternative user interface for users?**

Where conversations happen

Software vendors and open source communities are gearing up to deliver conversational interfaces to all software systems by enabling chatbots to talk with enterprise systems. This will allow workers to use the same conversational interface they use to interact with their colleagues to also interact with their HR, finance, sales and service management tools using words. The lines between systems and people will be further blurred as AI technologies augment human capabilities.

The surge in the adoption of new digital channels to support distributed work has been astronomical, and the marketplace for apps to support collaboration is incredibly crowded. You may have wondered why so many vendors have claimed a stake of this patch. It's commercial of course, with the answer being a long-term strategic play to own the platform that people work in. Vendors that offer the daily communication systems are trying to "build, buy or partner" traditional enterprise systems (including finance, HR, sales and service tools), while those providers are buddying up with collaboration systems to achieve the same breadth of capability. This is one reason why Salesforce.com recently acquired Slack. Slack could effectively serve as a conversational interface to all of their other solutions, and why Microsoft is working so hard to capture as many users as possible in Teams. The quest is to be able to offer it all, glued together by, and experienced within communication

CONVERSATIONAL BOTS

The mention of 'bots' elicits a wide set of responses from people. For some, bots are shrouded in mystery and belong in the dark web or in the movies - some kind of magical tech thing. The truth is they are simply software applications that emulate one or more aspects of human behaviour. Chat-bots that communicate via text are perhaps the most frequently encountered. Due to the widespread adoption of collaboration platforms, employees often interact asynchronously via messaging. The intelligence that bots display is growing quickly and It's feasible that at work, bots and people may become indistinguishable in behaviour. The foundations are now set for bots to play an important role in the workplace, including performing those functions that have traditionally been seen as management ones.

enables organisations to adapt people management practices to help achieve business goals whilst improving the employee experience. Management may further erode the need for dedicated middle managers.

- **What are the management tasks that can only be performed by a manager?**
- **How autonomously can your teams operate? How do you know?**
- **Are your teams getting the things that only human manager can provide from them?**

do what they need to do without a lot of oversight, and is the basis for appeals for "Bosses, get out of your employees way" (Sutton). Self-managing teams are now widely accepted as standard practice. At the most they need a team leader (that is also a proficient individual contributor) to help facilitate their internal management processes. The middle manager's role has had a big chunk taken out of it.

Along with these collaboration platforms, technology has changed to help teams execute on organisational strategy, by cascading strategy down into teams and ultimately tasks. This helps teams spend their time and energy on things that matter, drop those that don't, and helps the executive see progress against the strategy. Again, the middle manager's role has been eaten into, and 80% of workers claim that they would be able to effectively do their job without their manager (Bolden-Barret).

The management response to lock-down has led to a situation in which there is a diminished need for managers. Software applications properly used now make it relatively easy to bridge the gap between the executive and the front-line to ensure that strategy is executed. However, the role has not entirely disappeared and what remains is important.

Workers need motivating, coaching, reminding, and updating of context (especially as it evolves) and also, to be heard. Unfortunately, these kinds of interactions are not the kind that managers have been focussed on, and as previously established, managers are struggling to deliver on the more "human" requirements of their role.

The gap left vacant presents an enormous opportunity for the application of data-driven techniques such as conversational AI (bots) that can provide guidance 24x7, People Analytics that can collect and interpret employee voice, and Agile HR which

Bots are finding work

Bots are not magic but they are finding jobs alongside workers, and in some situations replacing them. This is true especially for roles in which people work like robots, but it is becoming more common in other areas as well.

Without bots, the world of sci-fi would be poorer. The relationship between bots and humans has been a mainstay of too many plot-lines, with bots wrestling with their identity as a prop to force their human cousins to do the same. Their presence also looms large in the popular media about 'the future of work.' Discussions with where they will fit with human workers inevitably draws upon such fiction tropes, and perhaps engenders a degree of fear as a result. So before going too deeply into the fascinating topic of bots at work and their role in a remote workforce, it is worth establishing a couple of more modest perspectives to demystify the 'bot'.

1. If you walk into the average workplace, you will see people behaving like bots. As they swipe through the front gates, their demeanour (and facial expressions) fix into place as they adopt their work persona in preparation for the day ahead. Bots aren't new; it's just that we have been turning humans into bots in the office. The turn on between the hours of 9 and 5, recharge their batteries from the vending machines, empty themselves of hope and aspiration, and follow the rou-

tine. People complain that their work requires them to behave like droids. Yes this is a glib and often unfair observation, but in this sense, bots at work are nothing new. It's just that humans have been volunteering to be meat machines for decades.

2. Secondly, and something that I still find amusing is that South Africans call their traffic lights at intersections robots. It is a good name. Where once stood white-gloved police officers now stands a set of colored lights. Developed in the '50s, they are highly resilient logic machines that communicate instructions. They do not pretend to be human, and lack knowledge of the protocols that C-3PO understands, but they fill a role that humans once did.

Bots need not be sophisticated AI machines. Bots are simply software applications that perform tasks automatically. In this sense most software applications could be considered bots at some level, but modern bots differ in that they also seek to act like people, communicating in natural language.

The marketing has been good. Many seem to think a bot is like what we see on star trek - a computer application with such broad general intelligence that it can go explore the world, acquiring experience and information independently. Maybe that is a future we should all be careful not to create and is the concern of the authors of the "Open Letter on Artificial Intelligence." Wherever they have come from, and whatever they are developing toward, bots are finding their way into the workplace and filling a number of important roles today, based on their existing capabilities.

A field guide to office bots

You will discover bots that emulate human behavior in some capacity working in a range of roles in offices.

Sales and marketing

The sales function was one of the first areas in businesses to use conversational style bots, appearing on websites to stream inbound calls or provide the first line of discussion with customers. Sales bots generally have a one-track mind to progress people through the pipeline. They may hand customers off to another person (acting as a buffer that streams customers to the right point), or they may attempt to complete a discussion. Sales-bots are designed with scale in mind, to create a cost-saving in comparison with fully human-based practices.

Service and Support

Hot on the heels from their sales-bot cousins, service-bots followed fast into contact centres to handle general inquiries and to deliver support to customers in need of help. As with sales, service-bots usually work alongside staff to help answer questions quickly or to multi-task conversations, to identify angry customers that need extra, and to capture successful resolutions to improve future conversations. The service-bot is often a way of delivering an FAQ as a conversation, a ticketing machine where issues are not resolved, or extended to perform basic common requests (think of the number of times a service desk does password resets after a holiday).

Human Resources

How ironic - services for humans who are described as 'resources' delivered by machines imitating people. Bots are finding a happy home in many corporate service teams and responding to common queries from users such as 'how much leave do I have,' 'I want to update my details.' Human Resource departments (along with finance, legal, property, and others) are finding ways of getting more done by handling common queries using bots, and using them to help communicate and instil an understanding of various policies and expectations into the workplace. The HR-bot is a popular application of the technology as it takes what is often highly detailed, and hard to find information directly into the tools that employees are using daily, making the policy burden more easy to handle (Bailie, 2019). More recently, HR-bots are also being used to listen to "employee voice" and gain feedback on specific topics too.

Intranet Replacements

Over time organisations have invested in intranet sites where essential information is meant to be accessible and where all-staff communications and announcements can be found. Intranets have mirrored the outside world of the internet but provide employees a change to provide a "walled" experience" with organisational branding. Except for rare exceptions such as GitLab, intranets have relied on teams to develop strategies, information architecture, visual cues and symbols to help users find the right information, and content moderation. The intreat-bot turns this inside-out. Rather than banners, headers, navigation bars, hierarchies, and buttons, the bot reads the information stores and serves the information up in conversational form. Why go to a browser and search for documents/pages that might contain an answer to what you need when you can just ask a bot in your communication tools and get an answer? Just like phoning-a-friend!

This inside-out approach to information delivery also establishes a basis for understanding the information needs of a workforce. Bots can not only serve relevant information but they can monitor the information access patterns (much more deeply than page counts), highlighting what people are keen to know, what they don't understand, and what information gaps need to be filled. This extra data create an opportunity to overcome the challenge that a rapidly stagnating collection of content creates, by helping define the content life cycle more dynamically. By 'training' a bot rather than publishing a document on the website, it shifts the focus to the person's actual need rather than the document as at the presumed solution.

A Pattern for Systems Integration

The conversational UI covered in the previous chapter allows people and applications to work together using only human language. The blend of natural language understanding, predictive recommendations, and robotic process automation will simplify the employee experience, reduce the learning curve of workers, and provide IT departments with a layer of abstraction that allows them too change underlying systems without impacting users. This final benefit can also extend the life of legacy systems. Adoption of the conversational UI is currently being led by IT departments (Schmidt), but mainstream consumer use of smart speakers provides a strong indicator that people (including workers) like to interact with computers without screens

In-channel

Bots are now 'living and breathing' within the channels that staff use for everyday communication. Popular systems that provide collaboration capabilities have extension points that allow these applications to be deployed as if they were members of your organisation and interact with staff in precisely the same chat and conversational ways, but backed with all the power of software systems. As bots take on tasks within organisations and develop individual personas within the social organisation, the sense that they are software applications will diminish over time. Instead of being viewed as a technology applications, they will come to be seen as members of the social enterprise (Ingham).

- **Where are bots currently working in your office?**
- **Can you see natural applications for the use of bots in your workplace?**
- **Do you think you could augment your workforce with some form of automation, and if so do you have a transition plan?**

Under the hood: how bots work

Bots are not magic. They are software applications that have interfaces that mimic human ways of communicating. Rather than relying on buttons and menus on screens, they work using natural language, physical gestures or other human cues. Knowing how they work, and what their capabilities are will help understand what will influence they will have in the workplaces of the future.

In Hollywood blockbusters, bots come in all shapes and sizes. They do in real-world offices too. Understanding their differences helps explains their diversity, current barriers to adoption, and their potential future applications. Knowing how bots work can help explain why they will play such a pivotal role in the future of work. The first and most obvious distinction between their types is how they communicate (or interface) with people.

Text bots

No doubt you have experienced the frustration of chatting with an unintelligent bot on a website that has popped up to 'help' you. Not only did it appear to be unable to deal with your obvious queries, it couldn't cope with anything complex, and couldn't stop repeating itself. Endless loop anyone? This is not the bot's fault. It has been left to do a job it hasn't been adequately trained for. People might complain about Siri's default fallback response; when it isn't sure what to do it turns your words into an internet search. For example, when I ask “Hey Siri, what is the name of the big scary dinosaur with short arms”, it responds with a list of websites and asks me to check them out, but if I say 'Hey Siri, please turn the music up' then hey presto. At least it knows what it doesn't know, and when it deviates from its 'happy path' it can do something half useful as a 'fallback' in response. That we feel disappointed probably points to our growing expectations (or rather our demands) of being understood. By the way I wanted it to say *Tyrannosaurus rex*.

Human communication is intricate and subtle. Computers find such nuance hard to mimic. One way of levelling the playing field and give computers a head-start is to remove the chance for eye contact, body language and tone. The most obvious way of doing this is through text - like SMS and email, where words can be exchanged in turns without having to cater for higher order signalling. The relative ease for machines to imitate humans this way was foreshadowed by Turing in his famous “Imitation Game” from 1950. It focussed on whether text generated by a machine was discernibly different from that of a human. Of course, this is exactly how a large proportion of workplace interactions happens these days in the office via their chat channels. Its a big reason why flame-wars happen without the nuance. The science and art of algorithms that handle text in human ways is progressing quickly, and for short exchanges, machines can already converse in ways that are not distinguishable from people.

Conversation Emulators

Conversations between friends are usually unconstrained and meander off into many different directions. In contrast, some types of conversations (like those that happen in call centres and sales channels) can be reasonably well scripted and repeated. These predictable, single-track conversations have only a fixed set of options or pathways to them, and can be visually represented as a simple flow diagram. 'Button-bots' that emulate conversations of this kind have no understanding of language, and do not learn, but simply offer a sequence of options/paths with text labels on buttons. This gives them the feel of a conversation, much in the same way that the 'Choose Your Own Adventure' books of my youth did. This approach is popular for bots on websites to get people from point A to B, and depending on how finely-tuned and complex the conversational paths are. I was recently 'interviewed' by a bot of this kind that was updating my records by SMS. Rather than giving me a form with text boxes, it just had a set of text messages. It was straightforward, but limited to a highly structured path.

Conversational emulators are simple as they require no real attempt to handle or interpret language. They make no use of AI or algorithms. Instead they just interpret a sequence of buttons like a vending machine and so cannot respond to any user needs that haven't been anticipated by the programmer in advance, nor can they self-adapt over time. The approach also becomes hideously complicated if the goal is to permit a broad range of conversational paths as every pathway must be explicitly defined by a developer, and the more 'networked' the conversational becomes, the less useful they become.

Natural Language Understanding

Natural Language Understanding (NLU) relies on parsing text and predicting what, based on historical examples, the most likely intent of the user is. NLU removes the complexity of having to map all possible user inputs to all possible outputs in a complex spaghetti of rules necessary with 'button-bots'. It does however, rely on a sophisticated data processing pipeline that groups similar words together, standardises terms, identifies entities, tags different parts of speech, highlights negative and positive sentiment, remembers details for future use, and gels all the details together in an algorithm. This training process creates a model that a bot can use to hold a live natural conversation within the sphere of knowledge in which it has been trained. Despite the sophistication, 'understanding' is an exaggeration as the process is just an intricate pipeline of functions and procedures applied to text.

NLU is a proven approach for task-oriented work such as "switch bathroom light on," "show me my leave balance," or "set a reminder for doing my expenditure report". It can support numerous tasks simultaneously and can maintain conversational context, remembering the entities that have been spoken of already. NLU has advanced to a state of maturity where it's now a relatively straightforward exercise to connect data pipelines and pre-trained models built by data scientists to software applications, and so has become mainstream.

Search-bots

An alternative but no less data-driven approach to holding conversations with users is to rely on 'search' algorithms. As with google search, in its most basic form this approach takes inputs from a user and calculates relevance of all potential responses, treating inputs as 'search' terms. Like NLU, this can also leverage a range of libraries and pipelines to extract entities, sentiment and standardise text, but it differs in that it doesn't require historical (or fictitious) conversational data to train the model with. This makes it useful for situations in which a vast array of different responses might be needed, where building a dataset of conversations is not feasible. User guides, policies and FAQs are popular sources of data for search-bots and despite the conceptual simplicity, they can also be surprisingly powerful.

Voice enabled

As your smart phones and speakers show, bots can also interface with humans via voice. Algorithms can learn individual voice signatures, detect sentiment and tone, and automatically transcribe sound waves into text. Contact centers are using algorithms of this kind to perform real-time routing and escalation of customer calls (Nichols, 2018, Shacklett, 2019). The effort involved and money invested in building voice assistants has been huge. At one stage, my Australian accent would throw these systems out (though not as frequently so as my Scottish friends), but they rarely make mistakes now. It's very impressive.

Text to speech to text capabilities are also readily available in browsers and other applications, originally designed to help with accessibility. Once sound waves can be translated to text, and vice-versa, the whole realm of text processing becomes available. This means that the bot can speak back to you, and hold the kind of intelligent conversation that can happen via text but with voice. As an extension to Turing's 'Imitation Game' in audio, I wonder whether Alexa could make it as a priest in a confession booth? Perhaps this is an option for a dwindling clergy. While voice is popular in consumer applications, it is much less commonly used in the workplace. It is easy to see though, how once coupled with corporate knowledge and access to workplace applications that it will become a mainstay.

Consider for example during a discussion with colleagues your very own Cortana-like bot saying:

- 'we've heard this one before, maybe you can keep this story short'

- 'you've been talking too much during this meeting why not let someone else have a go'
- 'no need to get angry'
- 'the financial results are not quite that good. If you look at .....'
- 'Actually the policy states that....'
- 'Why don't we try......'
- 'That document is way too long and hard to understand. Only 2% of our workforce will get it.

It's only a matter of time before there is a bot perched in every Board Room. Armed with the financial, contractual, market, and people data, with a flair for creativity and a thorough understanding of corporate law.

I wonder what their salary will be?

Visual gestures

If you are old enough, you will remember the Wii and its motion sensor handsets entering lounge rooms and being accidentally flung into TVs. Microsoft's Kinect 3D camera system was a breakthrough. They could locate people and movement without handsets (or headsets). TVs were heard worldwide to breathing a collective sigh of relief. Virtual and mixed reality headsets can also track micro gestures of users on their face and hands as they navigate virtual worlds and control the scenes and items in front on them. Machines have become adept at taking a range of physical movement data ranging from telemetry, location, and visual inputs to drive them. Facial recognition (and fingerprints) have also become popular for providing biometric security, and emotion detection, and it has been possible to search images from your browser, and locate photos by example for a few years now. While not new some of these innovations have been held back from wider adoption at work due to the clunky nature of wearables, network limitations, the legacy systems that they interface with, and the occasional bout of motion-sickness, but the ubiquity of cameras, increasing convenience of wearables, and ongoing increases to bandwidth with 5G will provide a basis for a resurgence in the prevalence of bots seeing and responding at work.

Robotic Process Automation

An automation technology called Robotic Process Automation (RPA) is usually what springs to mind when people think of bots at work. What started out as an approach to automate software testing by tracking and repeating tester actions on-screens, RPA applies the same technology to using live systems for actual system use. RPA relies on building a synthetic 'robot' user that can emulate their actions using screen layouts, information such as X and Y coordinates and pixel locations as deeper semantic identifiers encoded in the content of pages. RPA can perform actions such as clicks and data entry on those structures, much like a human user does. And just like a user, it needs to be trained, and this is often achieved by 'recording' the session of a human user well versed in using the application. RPA has become popular for extensive data entry tasks, in processes with manual 'bottleneck' steps, or where users need to access multiple systems to perform a workflow. While some up-front costs are involved in training the bot to press the buttons and access the screens, if the task is performed thousands of times per day, the return on investment is high. Of course if the screen changes, the bot might need re-training as they aren't intelligent enough (yet) to adapt the way a actual human user does with minor updates, but nothing is perfect.

RPA when combined with NLU, voice and other interpretive services to interact with specific enterprise software systems purely with natural human interactions such as text and voice.

Multi-sensory

There is no reason why bots cannot use inputs from all senses, combining text, visual cues, sound, and touch. At this stage, most bots at work are single sensory. This makes them one-dimensional and naturally lacking in the degree to which they can imitate humans.

Time will change this.

- **Does the idea that bots will have such a broad range of capabilities concern you, and if so, why?**
- **Do you think bots are a fad, or is there something you should do to prepare for their growth and adoption?**

The bots are getting better

Bots are becoming increasingly capable of the kinds of things that were previously seen as distinctively 'human'. This includes the ability to synthesise, analyse, reason and create. These are many of the things that we expect of management.

The capabilities of bots are rapidly progressing and by the time that you read this, they may read like ancient history.

Writing books and articles

In 2019 OpenAI announced GPT-2 (Generative Pre-Trained Transformer #2) to much fanfare ("Better Language Models and Their Implications"). Touted as "too dangerous to share" (Vincent) GPT-2, was an algorithm designed simply to predict what word would most likely come next after a series of other words. This seemingly simple task is in reality quite a challenge, but being able to do so means that a bot can generate its own paragraphs with just a little bit of a start, and the occasional extra bit of help. It received much attention at the time as it was able to generate an 'article' based on a brief set of seed words. The output appeared similar to what may have been produced by an imaginative human. The input involved unicorns, exploration, and South America. The output was a ripping yarn about the history of the exploration of new territories by a group of adventurers who discovered a previously hidden herd of unicorns in a hidden valley. A year later in 2020, GPT-3 was released, and if you believe the journalists, it wrote its own articles to announce itself - though the fine-print shows that it had at least bit of an editorial helping hand (GPT-3, 2020). It's a big step toward what is termed Artificial General Intelligence (AGI) whereby machines can apply intelligence gleaned from learning about specific problems, to other previously unseen "zero-shot" problems. This is like a human's ability to develop "rules of thumb" that can be applied in novel situations.

It is easy to see why reports claim that about half of the tweets about COVID-19 were made by bots during the US election, and why sports news is now often put together by bots "recapping" a game based upon its timeline and statistics. I wonder how long it will take until CEO, divisional or branch updates are provided using similar techniques, or creative agents complete weekly status reports. It may be enough just to prime the bot and let it go. The discussion of the update might make for some interesting 'all-hands' meetings.

Synthesis and summarisation

Search engines provided instant access to information from around the globe. However, what they provide to users are lists of links to pages for people to trawl through (though annotated summaries of the best results are getting better). Users still need to distil the salient points from amidst the noise, correlate information across separate documents/resources, and form an opinion.

Bots are now extending their utility by performing this activity on their behalf, through textual analysis of results, and synthesizing content from the links into digestible forms. A classic example of this in practice is in the legal world, where rulings, arguments, and publications get scanned by people for relevance, and a collection of them synthesised for detailed analysis. Legal clerks have done this work for many decades, but now they are either supported by, or replaced by bots (Akshara). The same can apply to competitive intelligence in marketing or media monitoring services. Where once sat people, reading, highlighting and tagging, bots now trawl content, identify relevant articles and entries automatically, and bring them together into thematic and topical threads. Summarising is perhaps the trickiest step in this process. Bots typically summarise by:

1. extractive summarisation - finding the snippets that best reflect the overall range of expressed information; and
2. abstractive summarisation - developing new content that captures the meaning in the information (really hard to do)

I wonder how long it take for bots to make a detailed analysis of your business and develop a strategy for it to follow?

Collection and Analysis

Bots have often played a technical role in monitoring systems and inventory, and alerting administrators to issues or diminishing supplies. As bots become more sophisticated, they are becoming more proficient with the collection of data from various sources, highlighting anomalies, and communicating potential issues, and there are predictions that we will see the widespread use of bots to independently exploit whatever data is at their digital fingertips by 2030. This is a natural extension of bots, connecting a conversational agent to the back-end data analysis systems. We are now witnessing the adoption of voice-enabled reporting tools interpreting dashboards in board rooms. While a picture tells a thousand words, and the voice delivery of insight may be less useful than a well-constructed visualisation, having a bot that can rapidly answer any question from organisational data on demand is likely to open up a wide range of alternative applications.

Discovery and information granularity

If you've ever had to perform an audit of user actions to supply computer-based evidence, you'll know this can be a time-consuming process. Stitching together and interpreting system logs can be hard work. Sometimes you also have no notion of what users were doing when an event such as a page display was rendered. Did they look at the screen or read it? It isn't possible to tell what happened to information once it left the screen. In contrast, conversational systems narrow the gap because an audit log looks like a conversation, and the data it holds is far more granular than a page view. It is possible to see specific information exchanged, and responded to, meaning that data must have passed the eyeballs, entered the brain, and trickled back out through the fingers onto a keyboard as a response. If bots are used as interfaces to back-office systems, then the discoverability of user actions will be simplified while becoming more powerful.

Creative agents and deepfakes

In certain contexts, bots need to be carefully controlled to restrict their behaviors to a particular predefined range of possibilities. There is no point in having a chatbot that is meant to explain the law that decides to get creative, and you do not want an online chatbot recommending that customers try competitor services. However, creativity is one key area of development in the field of bots. Engineering creative initiative Ito bots promises to make bots both more powerful, and better impersonators of humans.

I have already described the creative powers of GPT-3 as a method by which bots can build content (natural language generation). Another relatively new class of AI learning technique called Generative Adversarial Networks or GANs has emerged to solve what were previously unsolvable challenges (Rocca). GANs pit one neutral network against another in a contested arms-race, with one network trying to generate fake versions of something (e.g., pictures of human faces) while the other network attempts to distinguish between the fakes and real variants. The two networks are linked to share their findings, and then try again to outsmart one-another until they finally reach a point where fakes are indistinguishable from originals. The approach works well with anything that produces a storable digital asset, such as text documents, images, video, and sound. It is from the use of GANs that Deepfakes have appeared.

The deepfakes video phenomenon has risen to prominence with famous people being made to look like they are saying unexpected things. This has included advertising where newscasters from the past have made incredibly specific predictions about today's events, and more incendiary remarks from political leaders (e.g. Galston). There is no better example that Donald Trump's fictitious concession speech, which was clearly labelled as fake, but how would people respond if it wasn't labelled so? It's not a too big a stretch to see how the phenomenon of deepfakes might create a society in which citizens are unable to trust or distinguish truth and fiction, leading to societal breakdown and conflict. More startling is the fact that it is no longer purely science fiction to reproduce images that people see only in their mind's eye from the physical thought patterns emanated from their brains - awake or asleep (Huston). Is this the world of the thought police foretold by Orwell?

The field of generative visual art is populated by an eclectic mix of creative artists and computer scientists. For years, people have been pushing the boundaries of using technology in art, but recently the creative process has also been granted to computers. Initially geared for artistic purposes, computers were taught to create images with different criteria, styles, and moods. Such art is not limited to the abstract spectral screensavers of the 1990s. It has also been used to generate artistic works guided with input from people, though its not to everybody's taste (Elgammal). OpenAI has turned its attention to this challenge as well (Hao, July 2020) using the same unsupervised and human-unfiltered approach as that used for one-dimensional text in two-dimensional images, predicting the order of pixels, completing the second half of a picture. Not stopping there it has also released Jukebox.ai, a system able to generate music to your preference of genre or musical group, though I can spot the difference between an actual AC/DC song and an AI imitation. In homage to both Turing and the band, you might say it is not quite "Angus Young complete".

The ability of bots to move beyond the simple execution of processes to do higher order tasks, including those requiring creativity still lags a long way behind humans, but the gap is narrowing. Who knows how such advances may affect work, and what functions bots may play in the workplace of the future, but it seems very unlikely that their job descriptions will be limited to basic task automation tasks, and the bot glass ceiling is going to be broken.

- **If bots continue to increase their capabilities, what kind of impact do you think that they will have on the workplace?**
- **Are there any things that people do that you think a bot will never be able to do?**
- **Would you like to see a bot try a creative task?**

But bots can't climb stairs

Bots are not human, and despite the incredible progress, still have limitations that may never be overcome. However, as people's expectations increase and as the cost and complexity of tuning them decreases, they will start filling new roles in the office that require higher degrees of education and training, and do so very competently.

"Real Daleks don't climb stairs; they level the building" goes a popular joke amongst Dr. Who fans. For years, fans comically asserted that the Daleks quest to conquer the universe was constrained by their apparent need for a smooth floor for locomotion. It was only years later that the writers revealed them as able to overcome stairs by hovering and flying. Similarly, the relative maturity (or lack) of conversational agents at work has left many with the impression that they are mere faddish toys. This is unfortunate but understandable, as calling most of the bots unleashed today in the workplace incompetent is a charitable assessment.

Setting expectations

Some people think AI works supernaturally. They will always get disappointed by the reality. Making sure that people have reasonably set expectations is a good first step in avoiding disappointment. One way of doing this that I find helpful is to offer visual or verbal cues about the maturity of specific bots when people 'meet' them. Early releases of a bot might be considered analogous to a 'baby' that clearly shows the user that the bot has limited understanding and depends on its parents for training. Lo-fi logos, introductory welcome messages, and names that suggest immaturity are usually enough. People are often slightly reluctant to engage with 'baby-bots,' because they might be wasting their time, but the first set of interactions is useful as it creates more data to permit further development. A 'toddler-bot' on the other hand is eager to help, even when it isn't so helpful. It's clear that they have got some sense of what is going on, but they tend to get in the way. They can be cute to their parents, but toddler-bots are often annoying to busy adults. Some of the bots you have met on websites were probably 'toddler-bots.' With further education, they may develop into 'teen-bots' which resemble full-grown adults and may have mature arms and legs and able to complete tasks, but they are still learning to fit their skin, and their brains are still forming. Like human teens, their knowledge is drawn more from what they have read or heard rather than life experience, but when trained well, they can handle a lot of work given the right attitude. More advanced welcome messages, higher fidelity or sophisticated visuals, or prompts pre-empting user tend to signal this well. Eventually, if a bot survives long enough, the bot may develop into an ‘adult-bot.' By this time, it knows its way around, and how to help. They have mastered their area of expertise and can support user interactions and even adapt quite well in novel situations, taking the initiative and suggesting solutions. These can be rendered in full detail. Alternatively and possibly counter-intuitively, they can be made to just blend into the background and become an unseen but useful resource.

Symbols of what the bot is for and its maturity level may help users have expectations that don't overshoot its capabilities, leading to frustration. This may help provide bots the time to grow into maturing and become productive members of your social enterprise.

Complexity and cost

Like most technologies just entering the mainstream, the cost of deploying bots is in virtual free-fall. Until recently, baby and toddler-bots that perform simple tasks and fixed workflows have been expensive to acquire and train. In 2019 I set up a toddler-bot to help facilitate a conference. It knew the agenda, some assumed reading, collected participant feedback, and learned a little through simple instruction by participants. Deployed in 'Slack' (the social platform that attendees were using to stay connected and with one another) it was a useful experiment. The biggest surprise was not that that attendees loved it, and held conversations with it after-hours when back in their hotel rooms - it was the bill! I was slightly embarrassed that I wasn't more careful and I should have known better. Since that time, I use my custom-built bot to avoid such bill-shock.

At one point in time, training a bot required a software development environment and an engineer, data analytics environment equipped with a data scientist, and a project code to bill against. However, the costs associated with this are quickly declining and people can expend more effort in improving how they work. One emerging strategy to reduce the ongoing cost of training bots is harnessing real user interactions as part of an ongoing training process, so the bot learns live as it goes. Another is allowing the bot to acquire new knowledge directly through a conversation with people. In some situations, organisations will prefer that authorised 'teachers' centrally train bots. In other contexts, a bot or a family of bots might be raised by the organisational 'village,' and such training processes will facilitate a shared approach to this task, decreasing the complexity of bot training.

Lacking agency

One of the limitations of bots in the corporate context is that they operate passively, like a waiter, careful not to interrupt their clientele. Bots usually lack agency and are not encouraged to apply any initiative in their quiet moments. I see this as a lost opportunity. If I were hiring a waiter, I'd expect them to serve food and check patron enjoyment. If they also calculated which menu items were most profitable, built a roster based on peak times, set the music (and managed royalty payments), and predicted staffing needs over the next three months, I'd be impressed. I believe bots will develop agency and expand their usefulness.

In a modern enterprise, people commonly assume that bots will fulfil only the lowest level tasks. Bots are well suited to such pursuits, of course, but given that they can see, hear, research, communicate, and create we should expect over time that they will start filling, or augmenting higher order roles - such as middle management.

- **What is your experience at this point of interacting with bots?**
- **Have you been impressed or annoyed by them and what is it about them that has caused this?**
- **Do you feel that people are expecting too much of bots? Are they intimidated by them?**

Botty training

Like people, bots require continuous training. Once the remit of deeply technical data scientists and software engineers a shift is happening that is enabling bots to be trained just like any other member of staff. While a community of employees can take on the responsibility of teaching a bot, who is ultimately responsible for how they work?

"Who manages the bots?" is a question that will be asked more often as they become more widespread in businesses. Just like human workers, someone needs to "look after" them.

One thing that chatbots have in common is that they require training as they commence their journey and as they proceed through their life. Bots need structured data presented in specific formats to allow algorithms to build models of knowledge. The structures may include files of user inputs, parts of speech, tasks that bots should perform in certain conditions, stories of how inputs and outputs should be sequenced, or flows of dialogues defined in deterministic flow-charts. Depending on the need, bots require different levels of training. Up to this point, most useful bots have been specialised applications requiring a reasonable level of technical expertise to create, manage, and train. While several platforms allow for their simple creation, they have been the kind of naive 'button-bots' and 'Q&A' bots. Because the work is technical and expensive, 'ownership' of bots tends to sit with a centralised technical 'bot-team' that maintain bots in the context of projects. The team that supervises the performance of the non-humans is similar to how the HR department currently supervises employees (Griffin and Coleman).

However, increasingly sophisticated bots are becoming achievable with less sophisticated technical skills. Bots will be programmable, without enormous data acquisition and curation tasks, and without having hordes of data scientists cracking open their IDEs to build them. Instead, they will learn socially and in context, just as people do through conversation. As intranets were once the domain of technology folk, they are now used like simple word processors by all.

The skill involved in training bots in this scenario will be less about data science and AI and more about psychology and user experience. This change might lead to some variations on that model.

Specific persona representatives

Depending on the nature of the bot, and the activities they perform, they may need a specific persona to be accepted by their human colleagues. Will they be chatty, austere, comical, concise, verbose, deferential, or authoritative. There is little point in trying to make a bot that communicates mandatory rules chatty and fun. Likewise a bot designed to help people manage their wellbeing shouldn't come across like a prison-guard. It is now achievable to detect such characteristics in language, and we can expect text generation to have the same kind of abilities soon. In fact, individuals use language uniquely their own way. If bots are used as a 'proxy' substituting for people when they are unavailable it may be that best results will be achieved if the bot adopts a similar persona to their delegate, much in the same way that people are seeking to 'resurrect' the persona of their loved ones in digital form (Harbinja). Having to maintain bots with specific personas will confer some work on those who supervise them, the cost of which will need to be offset by the benefit of having their very own personalised if they are to gain traction. But as bot supervision becomes less technical, and the effort involved decreases we may expect individuals to take some pride in their own bots, giving them highly custom and personalised behaviour. This approach upends the general concept of centrally managed bots. Maybe the work of the future will be in educating a bot to perform duties on one's own behalf.

Distributed training

One of the more exciting developments in training bots is the capacity for bots to acquire information purely through conversations with a wide range of people. Just as it takes a village to raise a child, coordinator bots will be educated by the broader team, merely through conversation. The character of such bots will probably be general purpose in nature, and appeal to many, due to the fact that they have many masters and influences.

- **Are bots perceived as a predominantly technical capability that sits with IT ownership within your organisation?**
- **If individuals were free and able to create bots that enabled them to handle aspects of their job in different ways, do you think they would be open to doing so?**

PEOPLE ANALYTICS

From the early 20th century, business leaders have sought to understand the mechanics of work in order to gain a productive edge. As work evolved and 4th industrial revolution dawned on us, work became more tricky to measure. In tandem, the major cost and value of businesses become more firmly rooted in its talent, and how well they used their time. To gain insight, organisations became more reliant on the opinions of middle managers whom were well placed to observe front-line activity, and contextualise it within business strategy and goals. This was an imperfect compromise. Recent advances in People Analytics means that talent is now being quantified, tracked, measured and adjustments made. The data used is generated directly by staff in their work, leading to the emergence of more data-driven workplaces.

The emergence of 'People Analytics'

There has always been a desire to empirically understand the details of how an organisation's workforce contributes to its overall competitiveness, productivity and mission. This has been surprisingly difficult and made more so with the arrival of new models for work. People Analytics is a discipline that has emerged in response to this challenge, drawing on technical and analytical developments from a number of other areas, including but not limited to HR analytics.

Knowing the number of sales a business has made over time is relatively simple. Knowing the cost of manufacturing is also straightforward. Gauging the quality of the manufacturing process, or whether a major project is on track for delivery is not so hard, but understanding whether people are working "effectively" is incredibly difficult. For most businesses, how "Time, Talent and Energy" is spent is not only their biggest cost, but also the key determiner of their success (Makins and Garton). Domains such as finance have regulations that prescribe how data is to be used, but the treatment of data for the 'people' equation at work is more fluid. In part, this has been due to the constraints imposed by a lack of useful data and perceptions that it is unquantifiable, and seen as "soft". People Analytics has (re)emerged bringing the empirical basis back to work, and the application of algorithms that are the bedrock of AI has enabled a step-change in what can now be done. However, the evolution of People Analytics has been non-linear, and is not a story from the rudimentary past to modern sophistication.

Early development

During the industrial age, the role that people played in the system of production was directly observable. Just like the sprockets in the machines that people operated, human resource inputs could be quantified in a relatively straightforward manner. The number of people, their age, experience, the length of the workday, employees' skills, and work practices were relatively simple to measure, as was their impact on production. This simplicity made work amenable to study and experimentation. In the early 20th century, industrial work was decomposed and scrutinised in the famous time and motion studies of Taylor and the Gilbreths. Data actually drove work.

This era saw many experiments being undertaken. For example, a series of famous tests were conducted between 1924 and 1932 at the Hawthorne Works for electrical equipment in Illinois, USA. Changes to lighting, work hours, and break-times were manipulated in different factories to determine whether they had any impact on human production. While the results were inconclusive, and subsequent re-interpretation of the data has caused debate, what does appear to withstand analysis is that a psychological 'Hawthorne Effect' was at play, with a short-term increase in productivity being triggered but attributable to ‘being observed’.

This was a golden age of data-driven work, and fields like Operations Research revolutionised production. Optimization of production lines, simulating changes to supply chains, and sequencing of rosters meant that bottle-necks disappeared, quality improved, and wastage diminished. The Toyota Factory System became a benchmark and icon for this kind of activity. Since this time, data has continued to be used to varying degrees. In

"Strategic Human Resource Management, A Research Overview" (Storey *et al.*), summarise the history of evidence-based HR over the last 30 years, showing that data has been used in highly sophisticated ways throughout this time, but not universally so. Data-driven practices are not new, but nor are they ubiquitous.

Trouble

Work was becoming increasingly complex, and measures of 'productivity' more tenuous, as human inputs mapped less tangibly with business output. While factories still adorned the landscape, swarms of people started gathering in office towers, around whiteboards, with post-it-notes, and fixed in front of terminals. Their work involved the sharing of expert opinion, refining concepts, and innovating. Productivity was not a function of how much paper was shuffled. While the machines of factories could be laid out as a set of linear dependencies, office-work was less conducive to mathematical study, and businesses started to succeed or fail on the basis of what was going on in and between people's minds. The emerging competition for talent was exploding just at the same time that the visibility of individual contribution and performance was vanishing. With the lack of directly observable scientific data, measures (and management practices) became more intuitive, subjective, and less empirically based. The usefulness of simple people measurements waned.

Several strategies emerged to deal with this problem, each of which has become incorporated into the field of People Analytics.

i. Management reporting

Within an organisational hierarchy, it is the role of managers to act as links in the chain-of-command, to connect the bottom of an organisation to its top. For this to be effective, individual managers need to have have effective lines of communication with team members, and a responsive management up the line. Management reporting is used to aggregate and cascade business performance to support decision making by those with authority, but it relies on individual managers understanding their people and their work and being able to vet the details. We have seen previously in Section Two that this has become impossible, especially with remote work.

Other approaches outlined here overcome this challenge by opening a more direct line with staff through data that bypasses middle managers.

ii. Sentiment and Engagement

Into the gap created by the difficulty in quantifying work came a slew of surveys, polls, and focus groups, to quantify how talent 'felt', how satisfied they were, and where the team should go and have lunch on Friday. Quantitative became qualitative. Measures such as staff sentiment and employee engagement grew in popularity and supplemented easily countable measures like headcount and completion deliverables. What the workforce said about itself became the focus of measurement, and acted as a new lead indicator of future business value. Numerous studies have demonstrated a link between these measures, and overall business performance (Kahn, 1990) and they have been widely adopted, remaining popular today. However, their value is coming under increased scrutiny, and in some situations questioned (e.g. Capelli and Eldor, 2019). It's unclear for example what the measures really mean, how robust they are as staff become familiar with the methods used, how gameable and prone to manipulation they are, and how they can be practically used to further drive business performance.

iii. NPS

The Net Promoter Score (NPS) has revolutionised the measurement of performance. Developed for customer feedback, the NPS was designed to be the one simple measure that 'ruled them all' (Reichheld, 2003). It is remarkably simple to implement, requiring that customers answer only one question. Its calculation is also easy to do with a pocket calculator or spreadsheet. As its popularity grew, it was applied to other issues, including employee experience. It was a simple leap to reframe the question from 'would you recommend this product/company to a friend' to 'would you recommend working here to a friend'. In tandem, anonymous rating websites such as Glassdoor started collecting similar feedback. Depending on the organisation, negative reviews were dismissed by managers as vindictive hate-speech, and positive reviews welcomed as fair and balanced. The eNPS (employee NPS) is now sometimes cascaded through an organisation's hierarchy, measured at multiple layers, and used to drive initiatives to lift the score within flagging areas. Fix the NPS, and you fix your business, so the mantra goes.

iv. Goal tracking

An alternative to measuring the human contribution to the workplace is to ignore it altogether. After all, if the business is achieving its goals, why does it matter? By cascading target outputs, achievements, results and indicators through an organisation and interpreting and rewarding them at an individual level, an 'action plan' for performance can be established. Key Performance Indicators (KPIs) for teams and individuals using SMART criteria (that is Specific, Measurable, Achievable, Realistic, and Timely) dominate, and Doerr's (2018) Objective, Goals, Strategies, and Measures (OGSMs) and Objectives and Key Results (OKRs), seem to have replaced the Management by Objectives (MBOs) advocated by Drucker (2009). So many acronyms. Provided that a set of measures are well defined which cover the most important outcomes, and relate to one another in a hierarchical manner, then it should be possible to monitor progress and course correct to maximise the chances of overall business success. These approaches have the additional benefit of generating role clarity as workers interpret their contribution to the goals of others. However motivating, the problem remains whether the lower level metrics are 'right' and roll-up neatly into a hierarchical map of organisational outcomes. Challenges arise when outcomes are not being achieved and the framework gives little insight into what the root cause is, or how to improve the situation, in particular if the challenges are rooted in 'people' and based on dependencies that have not been identified in the plan.

v. Behavioural performance frameworks

Over the last decade numerous groups have sought to understand the critical ingredients of individual and team performance in the hope of establishing and coaching teams for increased performance. By analyzing a wide range of observable factors these frameworks distil the core ingredients of performance, to be applied elsewhere. There are literally hundreds of frameworks to choose from, and the rate of their publication is not abating. COVID itself has triggered a wave a new behavioural frameworks for remote and hybrid teams. This popularity is possibly because measures can be tracked longitudinally, and lend themselves not only to quantification, but also for coaching based on the results. The downside however is that they have been developed in specific situations that may not be applicable more broadly, where high-performance behaviour may take on a different complexion. The following examples illustrate the range of approaches, but there are literally hundreds of others (e.g. see Haak, 2020).

Three Es

A research group from MIT's Human Dynamics Laboratory analysed teams from various organisations and discovered that their communication patterns were key in driving their performance. They found the best teams communicated amongst themselves with 'Energy, Engagement, and Exploration' - a model in which each of these factors could be broken down into lower levels of detail that are directly observable (Pentland).

Teams and Team Players

In "The Ideal Team Player" we read that three key personal behaviors critical in making people work well in teams is that they are "Humble, Hungry, and Smart" (Lencioni). This research followed on from earlier work that focussed on what goes wrong in teams (Lencioni, The Five Dysfunctions of a Team). If each aspect of the model is measurable, then it should also be possible to improve.

Seven Ts

Drawing on a study of more than 300 teams from 50 organisations, the Korn Ferry Institute developed the T7 model, defining both the intrinsic as well as extrinsic factors that most shaped team success. The seven T's of performance in this model comprise: 'Team support from the organisation', 'Team Leader Fit', 'Thrust', 'Trust', 'Talent', 'Teaming', and 'Task'. This model is holistic and includes some context agnostic aspects, such as team 'fit'.

Re:Work

Google has researched the elements of effective leadership and of teamwork and published the results under the banner of Re:Work (Rozovsky). For teams, Google observed several driving factors including: 'Dependability' where team members get things done on time and meet expectations; 'Structure and clarity', with high performing teams having clear goals and well-defined roles; 'Meaning' in which the work has personal significance to each member; 'Impact' where the group believes their work is purposeful and positively impacts the greater good; and 'Psychological Safety', where people can speak and show vulnerability without fear of attack or being undermined. All attributes can conceivably be measured by means of a simple survey.

Six styles

Emotional Intelligence (EI) has been demonstrated to be a good predictor of leader performance, as well as the performance of the organisational unit that they represent (Goleman, 2005). Goleman's subsequent leadership model (Goleman, 2013) describes six distinctive leadership styles (Develops, Demands, Mobilises, Cares, Involves, Drives). Unlike the other frameworks mentioned, this framework identifies the suitability of styles for particular situations, rather than good/bad qualities, and is therefore coaching oriented.

Each of these frameworks offers a set of dimensions or behaviours that are observable and therefore measurable (even if qualitatively) over time. If implemented within an operational framework, and assuming that they generalise to other organisations, they offer an opportunity to measure and coach performance. However, if the frameworks don't generalise, or if the behavioural qualities are not causally rated to performance, they may be meaningless, distracting, and potentially damaging.

The state of play of people data

'Step' inside the 'board room' today and the board report will show you what measures are deemed to matter. Most of the numbers reported will be financial, where metrics are standardised in definition from one organisation to another. Metrics-wise, finance is a solved problem, and executives can move from one organisation to another without missing a beat. Revenue departments have become data-driven machines, and marketing funnels and sales forecasts are *de rigueur*. Supply chains are quantified within an inch of their lives, and service delivery gets tracked down to the nearest cent and second. Procurement is controlled in triplicate, and project governance attracts the full spectrum of multi-colored status reports.

Yet surprisingly, in knowledge-businesses there are few standards that provide deep insight into the biggest drivers of cost and value. Businesses compete on talent and over 90% of value of S&P 500 is now intangible assets. 'People are our greatest asset' is a widespread adage, and talent represents up to 70% of operating costs (Paycor) and yet **how** talent is being leveraged is left largely unmeasured, and therefore uncontrolled (Makins and Garton). What we often see instead are reams of pre-defined dashboards and reports filled with workforce data. It's not uncommon for larger organisations with HRIS systems to automatically generate hundreds of them covering everything from headcount, attrition, projected workforce shortages, absences, with every other countable sliced and diced by location, gender, role type, and tenure. These are the kind of measures that emerged to support HR management for manufacturing businesses - where value was correlated with hours worked. For larger enterprises, HR analytics teams may measure deeper talent acquisition statistics such as cost-per-hire, talent development metrics such as course completion rates, and staff engagement measures from a cultural perspective.

With talent reporting, metrics that can be counted easily are counted in the hope that they have significance. However, they usually have limited value and are often politely skipped over in executive meetings. On occasions when they are further drilled into, decisions are rarely made, or the stats are explained away, because in the minds of the non-HR professional, they don't seem to connect with organisational value. It's reporting by the kilo.

On a positive note however, executives are starting to expect the same degrees of quantification and analysis of business investment in talent as they do for any other part of their business, and are starting to build capabilities for this purpose. People Analytics teams are emerging, and starting to generate this kind of insight, and it is happening without the need of support or assistance from middle management.

- **How well is your organisation's investment in people and how they use their time, talent and energy measured?**
- **Do you have any metrics that you can rely on to tell you this answer?**
- **What data about your workforce can't you live without?**
- **What do you wish you could know about your workforce that you can't get an answer to?**
- **Do larger organisational decisions genuinely factor the the impact on employees' work and the downstream costs that might be incurred?**

The tsunami of data in the modern workplace

The workplace is awash with data. Every keystroke, movement and even inactivity is recorded in vast information logs that can be subjected to analysis. This data is a valuable but is still a largely untapped asset that can be exploited to help understand the workplace's investment in people's efforts.

"The world's most valuable resource is no longer oil, but data" declared The Economist in May 2017. Just ten years earlier, that statement would have sounded ridiculous. Today it is merely contentious. A competing claim that 'people are our biggest asset' is well known. Whichever is true, data about people is sure to matter. In fact it is of such importance that as much as 85% of US stock market value is driven by intangible assets, including intellectual property and brand (Bersin, 2019), both of which are increasingly dependent on people and data.

Throughout the late 1990s and early 2000s, my office was a blackberry, an incredibly heavy laptop, and a collection of external spinning hard disks. Not quite Arthur Miller's "Travelling Salesman," I was a traveling data analyst. Today we would call it a data scientist. My job was to help clients get the most out of their data by building models that could help them validate strategic goals or to automate decision making. The work straddled customer acquisition and retention, risk pricing for insurance, consumer credit, fraud detection of various kinds, supply chain logistics, and a range of other interesting business problems. No one was analysing their internal people data.

The kit I had was revolutionary because it unchained me from an office. Previously, this was impossible, but I could now produce whatever I needed, wherever I was. There were not many clients wanting this kind of service at the time, so travel was necessary. The only time I needed to be anywhere though was to physically collect datasets for analysis, to present results, to supply on-site support, and when things were going well, to negotiate a contract. I could spend late nights at home, in hotels, or still on-site with the customer, working through a complication that I had neglected to plan for.

Despite the marvellous tech, my set-up was not unique. I have often been an early-ish adopter, but being on the bleeding edge is something I am rarely accused of. I had no idea that this was just an early prototype version of what a remote workplace would be. Another thing that escaped my attention, even as a data analyst, was that as I worked I was also creating data as a by-product. As I was analyzing buying trends, security incidents, and creating risk models, I was accidentally building a dataset about me and my work practices. This dataset could, in the right (or wrong) hands be coerced to reveal where I was located, what time I was working, the duration and peak time of work, how many times I attempted specific tasks and made mistakes, my contact list, and other details, constrained only by imagination.

It was only years later, as I was working in a large data-centric multi-national that the thought occurred to me, "why do we know more about our customers than we do about our staff." I had seen the power of analyzing social networks, predicting cyber-risks, longitudinal customer sentiments, and the like. However, I had never actually seen that analysis performed within an organisation to work out what made the staff tick, identify who was working with and against one another, and to drive better management practices. Business outcomes such as revenues and costs of work were always reported, but the data that sat behind those results, that could indicate it early, and that can be used to maximise the chances of success were not.

Wind forward to today, and you've probably provided your own staff with all of the tech tools necessary to allow them to work from places other than under the one roof. They can chat, video, and voice call, work on files together, allocate and track task completion, connect with customers, and perform a whole swag of corporate functions wherever they are. As they do so, they create a giant and potentially valuable footprint of data that could be coerced into revealing valuable secrets.

- **Do you know your customers better than your people?**
- **Do you know what data is created at work, and what value it might be able to add?**
- **When was the last time you looked at this data or imagined how it might be used?**
- **Who has access to the rich data sets being created by your workers as they do their jobs?**

New forms of data

Different forms of data are created at work including data that is actively sought from people, data that is passively sensed as a byproduct of their work, and additive data that combines elements of both. Each has merits. How data is sourced is an important consideration in determining how it should be applied.

Over the last decade, new forms of data have been exploited by analysts, leading to a heightened maturity level for People Analytics. When first described by Waber, "People Analytics" had a relatively narrow scope. It incorporated the analysis of data generated by people as a by-product of their work, including physical movement, machine and system logs, and communications data. This data is now often referred to as 'behavioural' data. The initial concept was to get behind what people said and felt to obtain an unobtrusive and unbiased method of getting to the reality of what they 'did'. Doing so, it was hoped, could help uncover the missing jigsaw piece of the data-driven workplace. No longer would leaders need to rely exclusively on the opinions of middle managers, or on surveys of the front-line to understand the state of the essential 'people' part of their business, but could see directly what they wanted to know.

Since that time, People Analytics has become a catch-all for the range of analyses associated with the human aspects of work.

Some describe this broad approach as 'employee sensing' (borrowing from the Internet of Things metaphor) or 'employee listening' (Coolen and Sombezki, 2019). These new capabilities elevate People Analytics to a more strategic role, with Chief People Officers now regarding it as one of the core competencies of their business (Guenole et al., Isson and Harriott, Bersin 2017). To understand why, it is worth looking at precisely what this new data is, and opportunities for its use.

Ask, seek and knock

Two terms that you will encounter when you start exploring People Analytics are *Active* and *Passive* Data Collection.

Active Data (ask)

Active data is harvested through an overt collection process, requiring participation from workers. It requires an *ask*, along with a volunteered *response*. Methods such as surveys, polls, interviews, workshops, and the like fit in this category. It is easy to obtain and depending on how it is collected more or less easy to interpret. Despite its simplicity it remains the 'gold standard' for measuring some aspects of the workplace such as staff engagement. Likewise, because it is data that is actively volunteered, the datasets are relatively small, and so (at least when structured in a survey as discrete values) it requires few or no specialist tools or algorithms to interpret. It is also a transparent process for staff, and provides them with an opportunity to provide explicit input (Judd et al.) and can be more efficient than workshops or meetings to discern 'employee voice'. These aspects reduce any potential for staff to feel like their opinions don't matter, or that they are being unduly scrutinised. On the downside, there are appeals to get rid of surveys (Cappelli) which are often distrusted (Wilkie). Moreover, employees can be swamped by requests for feedback, and can become fatigued and not respond truthfully, and with bias. In response to these issues, survey systems are being designed to be smarter and less intrusive (Bersin 2020).

Passive Data (seek)

To more fully counter the challenges of active data collection, some have turned to passive data collection. Passive data is generated in the process of doing work, and is collected in the background as a by-product. It is behavioural and is generated by worker activity. No *ask* is necessary as the data already exists, and only a *seek* is required. No extra effort on the part of workers to create data is needed, and is therefore not at all disruptive to their work. It is also potentially seen as more objective (Mattingly *et al.*) as it is not opinion-based, though this does not recognise that most advanced AI systems are now shown to carry bias (Hao).

Tech giants have been vacuuming up consumer data, recycling it into insight to be sold to third parties, but consider now the data that your people produce while working in the office or even more so remotely. There are streams of audio discussion over VOIP, video chats, phone accelerometer movement, location and proximity data, network access logs, web history, app usage, keystrokes, mouse movements, document accesses, eye-ball movements. More recently wearable devices, a market estimated to reach $135bn globally in devices alone by 2025 have started measuring electrical dermal activity, skin temperature, pulse and blood pressure so that sickness, stress, excitement and wellbeing can be monitored (Fickman).

It's the new gold, or oil, or bacon depending on your preferred metaphor, and as the available range of passively sensed data about people grows, what can be measured with it also expands. As an example in Project Tesserae the research team has followed workers using location beacons, smartwatches, and phone apps to understand workers activity. The aspiration was that by doing so, this passively sensed data might replace other 'active' performance measures such as manager observation and work review. Amazon famously applies these techniques in their mega-fulfilment warehouses, where there are clear linkages between the physical things people do, and their workplace performance. How these approaches relate to knowledge work is not so clear.....yet.

However there are some serious downsides with passively-sensed data collection, and especially with its interpretation.

- Firstly, the sophistication and cost of data collection and analysis is high. Thankfully, as we shall see in later chapters, new methods of data exploitation spurred on by advances in AI are making this task far more achievable than it used to be.
- Secondly, and more importantly is that there are limitations about what passive data can reveal. The Tesserae Study for example, revealed that frequent use of a phone at work was negatively associated with job performance. However, this finding was dependent on what the phone was used for (Mirjafari *et al.*) The trouble is, determining the type of usage from passive collection would require an algorithm to predict the purpose of calls, presumably by tapping into the the content of conversation of contact list.
- Thirdly, while powerful and achievable, such an act would likely have harmful impacts the office environment. I shall return to this topic in detail in the

chapter 'Building and eroding trust', but a common thread of using passive data analysis is that its interpretation and application is easy to get wrong without further corroborating data. Pursuing this track becomes an endless pursuit of surveillance, and its often simpler and more respectful to just *ask*. Depending on your view of what workers should be entitled to, data-sensing of this kind runs counter to regulatory trends for personal privacy, and while employees are not protected to the same degree as consumers, the trend toward privacy is strong.

Additive Data (knock)

Active and passive data are generally treated as discrete activities, generating separate datasets and at odds. However they can be joined together for richer analysis (e.g. Swain et el 2019). It is also feasible to merge them into a single 'process' that blends their relative strengths and weaknesses, into a single *knock* step. Take the following scenario for example. An instant message from a designated leader is sent to a team member that they do not frequently communicate with. An algorithm detects this novel event, and it triggers a specific request to the team member to provide feedback about one specific and observable behavior of that leader. The contextually relevant *ask* moment is triggered by a specific *seek* event in real time, to provide a valuable data point collected in context and with minimal disturbance, and because it is overt, it runs little risk of being perceived of as surveillance.

Rather than passive or active, I refer to this method of collection as 'additive'. The complexity in orchestrating of additive collection in a real-time requires a degree of intelligence (artificial or otherwise) to work out who is best-placed to provide what data, to give maximum insight without getting in the way of employees, but it offers the promise of giving the best of all worlds. Much like digital marketing analytics, additive data relies on algorithms, not just to interpret results, but to determine the 'who, when, where, how' of feedback.

- **Have you optimised any processes using this new form of data?**
- **Are surveys or perceived as a problem by your workforce?**
- **Have you looked for different ways of accessing and interpreting 'employee voice'?**

People Analytics in action

People Analytics is revealing a diverse collection of insight. It links how people and teams operate with organisational outcomes in novel ways, and demonstrates the benefits of learning from organisational data to increase business competitiveness and outcomes.

If the executive of a business could really understand how the investment in Time, Talent and Energy (Makins and Garton) was being used, they would have incredible opportunities to help drive the business forward, eliminating waste and lifting the quality of work. With new forms of data, and advances in AI, come new types of analyses. New insight has become available without relying on management opinion, nor on relying on a patchwork of non-consequential reports and dashboards. In this chapter I provide a small set of case studies that illustrate the breadth of insight achievable. If interested, there are a rapidly growing collection of books and blogs that detail a wide range and increasingly diverse examples (for example Guenole et al., myHRfuture). The corpus of such references will inevitably grow.

Organisation Network Analysis (ONA) with communication metadata

The analysis of online social media data is a burgeoning field (Martin & Hutchinson, 2020), with the analysis of links between people proving valuable for marketing purposes. The same can apply within organisations. As it turns out, behind the designed organisation is an informal network that reflects its actual work patterns (Waber). This network can be visualised by the analysis of the links between people, recorded as the shared use of documents, common team assignments, intranet page edits, physical proximity sensors, and messaging through communication channels.

Just like with social media, analysis of this data can be used to identify organisational influencers and top talent (Zielinski, 2018), and dynamics such as a more siloed approach to teamwork during COVID-19 lockdowns (Arkcol 2020). Being able to see how an enterprise evolves over time, without having to ask anyone's opinion is a powerful capability. It can help with managing the culture of an organisation, and facilitate structural re-design. Such changes typically take a long time and during the process a lot of water (and money) can flow under the bridge before discovering whether an intended transformation has been effective or not. An application of this approach has been described by Kienbaum where a a program designed to flatten the hierarchy and bring the best people to specific work tasks was run. Rather than using a range of project reports based on the opinions of managers, collaboration measures were tracked throughout the process from passively collected data (Leusch, 2019). This demonstrated a 50% increase in collaboration between teams, revealing that silos between teams were breaking down at various points of time. Because the analysis was continuous it provided an opportunity to course correct before it was too late.

At an individual level, because behaviour itself is measured, ONA can also be applied to create a template for defining which behaviours work best for what roles. With this information, business leaders can be much better equipped to design and tailor development programs for employees, and create long term succession planning using an empirical basis. For example, I have personally performed analysis of email metadata (not its content) to reveal:

- whether individuals were more extroverted or introverted along with the rest of the other Big Five personality traits with about 90% accuracy
- which people were most likely to apply for a stress claim is in the coming month
- whether an individual's opinion and knowledge was valued and sought by their peers
- whether individuals were decision-makers, influencers, stimulators or contributors
- whether a person was a natural leader of a business area
- if a person was a graduate, whether they were likely to be regarded as a high performer at the end of the year
- if a worker was a mid-level manager, whether they were ready for promotion.
- whether the workplace was split along socio-demographic or gender lines, and whether it was inclusive.

ONA has been ‘hot’ for a few years already, but one of its limitations is whether strong connections indicate health, conflict, or function. Metadata itself does not appear to be useful in inferring the polarity of any particular relationship, and at times, it is necessary for people to be unpopular to create change. Without context, it is easy to mis-diagnose individual relationships and act on the basis of incorrect interpretation. Instead, without additive data it is more useful for monitoring trends, developing aggregate views, and developing hypotheses.

Sentiment and emotion measurement

Unsurprisingly, the analysis of the content of communications between individuals can be valuable for gaining insight. The police have known this for years and rely on 'wiretapping' for investigations. But did you know, when you call a contact centre, the tone of your voice might help route it to staff with the right experience, especially if you are angry? (Nichols). There is already a range of commercial services that enable automation, using artificially intelligent algorithms operating on voice stream data. Likewise, text analysis is easily achievable from email and other text-based messaging systems and employee surveys. However, the practical experience along with the research reveals that today's algorithms are easily tricked by sarcasm (Berasategi). Leveraging this data will allow the executive, for example, to move beyond knowing who is rude, angry, or happy, toward more macro level questions such as what the workforce thinks needs improving, and how they feel about proposed changes. In a distributed environment where face-to-face contact is reduced, it can be difficult to verify hypotheses and create rapid feedback loops with employees without such algorithms.

Predicting performance with social media data

It's not only data inside an organisation that can be useful. For example, apart from physically scanning social media posts for vetting purposes, it's now also possible (with a reasonable degree of accuracy) to automatically extract social data to determine whether a candidate is likely to perform well in specific roles. Such algorithms make inferences based upon the skills described, how they respond online to other people, and the geographic spread and size of their social networks. Work in this area has also been extended to help identify whether discrepancies in how people present their work match the actual job description to gauge 'job ambiguity' and its impacts on employee well-being and performance (Saha, et al.) It's no wonder recruiting techniques have been moving away from primarily assessing CVs (Curtin). External social data has also proven useful for predicting resignation likelihood (de Jesus et al.), and no, it's not as simple as being alerted to a change in availability status.

The physical environment and performance

Analysis can be extended from people themselves to the physical environment in which they work. For example, in something reminiscent of the 'time and motion' studies of last century, synchronised lunch breaks for team members in a particular contact was demonstrated to be vitally important to performance (Waber). Who would have guessed that changing bigger tables so people could sit together during a coordinated breaks would have had a bigger impact on retention than pay and conditions? More surprisingly, in some situations providing staff with 'treadmill workstations' has been shown to increase work performance (Ben-Ner et al., 2014)!

As we come to grips with physical changes in how we work, especially remotely, this style of analysis will help discover the best ways of managing the workforce. Over the months and years ahead we are likely to see a lot of research on the quality of 'home' environments and the impact it has on business outcomes.

Right-sizing teams

Google has conducted People Analytics research for many years, identifying the attributes of their effective leaders and their most effective teams, finding that their best teams were comprised of between 6 and 8 people. This research accords well with the 'cost-of-communication' analysis outlined by Brooks' Law (Brooks) and the 'two pizza' rule of Jeff Bezos, stipulating that meetings - when necessary - should only have enough people in attendance to be fed by two pizzas (Cain). In the context of distributed work, where people can self-select in and out of ephemeral teams, keeping track of performance and team size may help work out the best division of labor, and the size of funding to particular team and programs of work.

Nudging for peace with feedback

In Work Rules!, Laszlo Bock, Google's former Head of People Operations, shares the details of a simple nudge he delivered. It involved a basic two-question survey that asked team members to evaluate the help they had received and provided one-another, reminiscent of the behaviours to reward and punish posed by Morieux in "6 Rules to Simplify". By compiling and ranking results, team members could see how they were perceived relative to others in the group. Interestingly, with no further intervention, the team perceptions of one-another shifted from 70% favorable at the outset to 90%. Like the Hawthorne Effect, the act of the analysis and the reminder of what characteristics mattered, couched in social comparison created change. A very inexpensive and effective intervention, even if it only lasted during the period of analysis.

A culture of risk

Catastrophic business failure is something we see in the news periodically, keeping CEOs proverbially awake at night. One of the underlying drivers commonly associated with these failures is a culture of deception, breaking the rules, and toleration of low integrity amongst the executive. It's no surprise that people have attempted to use analytics to help identify when this might be the situation, and surveys have been used for years to gauge organisations' risk cultures (Sheedy and Griffith, 2014). More recently, attention has turned to processes of 'employee listening' using passive data collection and analysis techniques such as text analysis of the words used in filings (Gupta and Owusu, 2020). Despite more work being required to validate this analysis, this approach would mean that tools are available to assess risk at overall and departmental levels within organisations continuously. This approach could continuously identify business areas that are not aligned with the overarching organisational risk appetite. It's simple to see how such measures could be tracked in real-time and reported in an ongoing way as part of overall business performance, rather than relying on personal observations of middle management (who have personal incentive to make their teams look good) and the occasional personal attestation.

Experience not required

When a business has hundreds of stores serving up food to hungry customers, it also tends to have thousands of temporary staff. Maintaining customer satisfaction requires quality service, and so it is essential to know what qualities to hire for, and how to develop new recruits. At a recognisable fast-food franchise analysis of a range of existing staff debunked several of the firmly held beliefs of its management team (Arellano et al.). It turned out that as far as performance went, experience on-the-job counted for little. What mattered more was personality. This finding can be simply put into action with a change in recruitment practices. What also mattered for restaurant performance was having managers who were willing to do the work and help in a 'hands-on' manner. It also showed that staff development was a more successful tactic at retention than providing variable pay incentives (such as overtime), and that rostering for a shorter than previously assumed number of hours was necessary to maintain peak performance and affordability.

It would have been easy to stick with the management team's incorrect expertise, but the data revealed another story, which led to serious business benefits.

Love your back

Workplace accidents are costly for businesses, employees, and their communities. Analysis of telemetry data from wearable devices has been used to accurately measure the risk of musculoskeletal injuries of individual workers. Further, personalised and coachable insights can be delivered directly to employees in real-time to avoid injury. This approach when applied amongst

450 workers has been shown to reduce the number of manual handling injuries by 55% over the course of a year. Using the People Analytics powered wearables, real-time alerts and suggestions led to significant costs savings, as well as an increased sense of well-being and trust by staff (SoterAnalytics). Relying instead on general communications, delivered as posters, and introductory training only yielded a fraction of this impact.

People Analytics and AI

People Analytics is simple in its essence and its adoption is on the rise. Its implementation boils down to using available data wherever it is sourced from to stop repeating mistakes, discover what works, and then repeat the things that work better, continuously challenging the *status quo*. Algorithms are able to see patterns that people can't, and it can do so in a much broader, more detailed and potentially more objective manner. While the implementation of insight can sometimes be difficult, getting direct, immediate and objective insight directly from the data to support important strategic and management decisions will transform the management practices. This is a better way of stewarding an businesses biggest investment than relying on the observation and expertise of middle managers.

- **Are there any mistakes or errors that are continually repeated in your organisation? Why?**
- **Have you crystallised why certain practices seem to work or have a method of repeating success?**
- **Do you or your organisation cling to any sacred cows or dogma that you suspect might not be right?**
- **What data analysis initiatives do you have underway right now focused on discovering areas for business improvement.**

People Analytics inaction

Despite the potential, it has taken longer than expected for People Analytics to make a real impact, and there are a number of reasons for this. Knowing why can help practitioners of the data-driven enterprise avoid the same mistakes.

The "Geeks Arrive in HR" announced Josh Bersin in early 2015, describing the changes that had been underway for some time in HR departments. At the time I took it as a sign that a wave was finally about to break, and I was ready to play my part. Yet in the same year, Rasmussen & Ulrich, wrote a paper to explain why People Analytics was not in fact a fad. This headline would hardly have been necessary if there was runaway adoption and success. Then in July 2020, the Economist published an article highlighting the 'new' role of People Analytics in helping organisations survive through the crisis of the pandemic. For five years it had felt as though the forecasts were remaining just that. Something with such an overwhelmingly positive proposition had stalled. But, the long-held belief of those involved had come finally to fruition, and the global pandemic and shift to remote work helped catalyze the change. Nonetheless, some of the pre-existing barriers to adoption remain, and there are still plenty of unsolved problems.

The immeasurable

Over the last twenty years, business analytics has surged, and has been adopted in waves through different parts of businesses, one department after the other. My observation is that adoption has been the simplest when analytics can directly improve existing measures of business value, especially when they appear in business plans or annual reports. For example lifetime customer value is directly influenced by customer churn, which can be predicted and acted on pre-emptively. In contrast the link between living out business values and the business outcomes that they are geared toward is often indirect and lagging. While People Analytics emerged partly in response to the lack of meaningful and practically useful people measures, the desire to find the 'one true metric' to put in the annual report remains a barrier to its further adoption.

Insight itself is not enough

Insight is interesting, but enterprises need to be able to operationally implement it or adoption will stall (Coolen and van der Brink 2018). After all, no one invests in the analytics for results to sit interestingly on the shelf. In finance, what to do in response to an overspend is straightforward - look for something to cut or another way to carry the cost. In contrast, when it comes to the workforce, the range of possible actions that follow insight can be overwhelming, and the best course of action is not often obvious. For example, assuming a sophisticated AI system identifies a staff member with a high risk of resignation within one month, what would you do? Celebrate, commiserate, offer more money or development, initiate a discussion, or wait and see? What would their peers think? Context is everything and if there are no obvious actions that flow from insight, having it is of no practical value.

When faced with this dilemma, the most algorithm-loving analysts would be tempted to add more sophistication to the analysis, and move from predictive (what might happen) to prescriptive (what should we do) analytics. In the context of people analytics the necessary data to support that work rarely exists. As a consultant, I have been guilty of presenting what I considered irresistible facts about an organisation, only to be surprised with the demoralizing shrug of "that's interesting but what can we do about it." Sophistication it seems is often over-rated. As Grant (2019) argues, it is sometimes preferable to quantify the obvious first, as it builds trust at least.

Thankfully the focus amongst practitioners is shifting more explicitly toward 'business value creation' rather than 'HR insight'. Frameworks such as that presented at Figure 1 offer an approach to reinforce this. A value focus will see analytical challenges framed differently, with the actions in mind. For example attempts to predict attrition (which would be an obvious task for someone with a background in customer analytics) might better be framed as learning what the causes of departure and the reasons for staying are amongst those considered great talent. This analysis would have more explanatory power to influence a positive outcome. Likewise, building a network graph of an organization to understand where collaboration is strongest and weakest may yield some fascinating results, but understanding what are the inhibitors of collaboration might offer some more obvious actions.

- **Can you actually act on newly generated 'people' insight with your workforce or is that a 'no-go' zone?**
- **Do you expect decisions about how management should be practiced to have as much justification as decisions about other expenditures?**

Gaming the metrics

An important benefit of funnelling work through digital tools of the kind that support remote work (including task management and ticketing systems) is that work becomes visible and trackable, obviating the need for triangulating whether people are working or not from other 'activity' data. On the surface, such data seems perfect for measuring individual 'performance', and it becomes tempting to start using measures like the percentage of tasks completed on time (a measure of deviation from a plan) as a measure of productivity. Unfortunately, measures of this kind are as much a measure of forecast accuracy as they are of delivery performance. They are also gameable. If instituted, one way for teams to make sure that there is low deviance (high performance) is to build a lot of 'fat' in. This results in teams comfortably cruising, and due to work expanding to fill the time available (Parkinson's Law), the fat gets used up, scope and/or quality end up being cut, and new estimates get created. Worse still, projects can fabricate progress by splitting, merging and de-scoping tasks. Other more abstract project measures, such as 'velocity' based upon arbitrary concepts like story points, T-shirt size, thrown estimates, planning poker, and the like, can be further manipulated to create a positive appearance. These measures were designed and intended for teams to use honestly to help them manage themselves, but issuing penalties and rewards on the basis of these measures leads to an unwillingness to commit to realistic approximations, overly-detailed up-front planning (which often stymies innovation) or manipulation of measures.

Other metrics are gameable too, and especially those where people volunteer opinions about others. For example, when engagement scores are broken down by teams, and managers rewarded (or otherwise), a common complaint is that managers make it clear to their staff what 'result' is expected. This dynamic is accentuated without guaranteed anonymity.

Getting the metrics of the analysis right is key in the successful application of People Analytics.

Limited Resources and Capability

Surveys of HR leaders often show that they encounter a skills shortage when setting up People Analytics capabilities (Green, Jan 2020). While data analysts are not so rare these days, finding people that can also combine analytics expertise with an understanding of what questions need answering, are able to influence leaders across an organisation, are able themselves to lead, and to work with new data sets is challenging. This 'skills shortage' is compounded as People Analytics is still awaiting a standard 'body of knowledge' to emerge.

An additional 'capability' factor is that the tools necessary for People Analytics are new, at least in comparison with other business areas. Platforms for well-codified analytic functions such as finance, marketing, and credit risk exist. However, the development of dedicated tools that make the full range of People Analytics tasks feasible are yet to mature. Organisations can therefore struggle to build a business case for the requisite investment in suitable technology, thought low-cost tools can reduce this barrier if adequate skills are available (Lupshor, 2010).

The skills and technology gaps have been a bit of a handbrake on widespread adoption, but these gaps are quickly being filled.

- **Do you have the tools or technology and the kinds of skills that will allow for analysis of the critical drivers of value from your people?**

Data Accessibility

When most businesses start with 'People Analytics,' they tend to start with data in their HR system. This launch point is a narrow view of what People Analytics is, but given the marketing associated with HR systems, it's no wonder that many go down this path. Sometimes they stick with data from these systems, confusing People Analytics with a kind of advanced version of HR reporting. Earlier I explained how People Analytics grew not just from data about people, but from data created by people as they work - behavioral data. Accessing this data can be hard, though internal security and audit systems are often an ideal source of data to mine. An informative example of this approach is provided from work at Westpac Bank (Dellala). In this situation, a specialised predictive data platform was established by drawing on a range of data sources built for other purposes. Behavioural 'event-based' data is often stored chronologically as a log, which is a highly flexible structure that can be used to understand the factors and causes of any kind of event that happens over time. By configuring this data for People Analytics, an analytics structure useful for a wide range of predictive People Analytics tasks was constructed.

As an aide, when investments Ibn software systems that people use for their work are considered, only rarely is the ability to access the generated data considered. Due to their simplicity of deployment and administration, the shift to remote work naturally fostered the adoption of cloud-based work systems. The flexibility they usually supports remote work as they can be accessed from anywhere with internet access, including BYOD. However, many of those systems were designed primarily for user adoption, and not to support data access. As a result, organisations can lose access to critical behavioral data. What is needed is to connect the data from multiple systems together, but providers will typically supply a fixed range of 'reports' that can be exported or in-built, standalone analytics. This does not enable a joined-up analysis of the workforce. It is also not always in the vendors interest to permit access to your data flexibly as it would simplify the task of migration away from their platform at a later stage. Care should therefore be taken when evaluating enterprise software to ensure that all data generated by it is easily accessible, in bulk and in real time.

- **Does your organisation have access to the data it creates as it does work, or does that sit out of reach**
- **Do you have data systems that relate to your employees that cannot be connected with other data for analysis?**

- **Have you tried integrating your various data sources that may shine a light on how well your people investment is working, or are you relying on a set of independent reports and out of date metrics?**
- **When confronted with a question about the workforce, do you sometimes hear that your HR system doesn't have that data, or support its analysis?**
- **Have your teams been seeking balanced investment across skills, analysis tools, and systems, or just in individual transactional HR systems?**
- **Is your workforce analytics capability funded to the same degree that other analysis functions are? If not, why not?**

Predicted Return on Investment

In a classic chicken-and-egg problem, it can be hard to make a compelling and accurate People Analytics business case before it is actually up and running. The costs are easy to work out but the return is much harder.

Unless there is a culture or framework in place that already values analytics, understands the necessity of approximation and permits limited investment in proofs-of-concept it can be difficult to just get started. By bundling data analytics in and other HR initiatives (such as measuring the outcomes of an L&D activity), many organisations have been able to co-justify People Analytics investments (Bersin, 2017).

- **Do you have a culture of testing and seeing what is possible?**
- **Has the case already been made that how people work is a critical success factor, and should therefore be analysed?**
- **Does all the investment in analytics get diverted to other parts of the business?**

Privacy and Trust

Privacy and trust is a vital topic that has attracted much attention in recent times. Frameworks for building trust have been proposed, but the discussion about the ethical use of data and AI has by no means been resolved yet. At one end of the spectrum, we have otherwise sensible people willing to put organisational data to work with no restrictions - apart from legal ones. At the other end, we have otherwise sensible people that are so concerned with the risk of using data and alienating people along the way that they shun any association with it.

The risk of getting this balancing act wrong, and being data-driven at the expense of all else is large, so I devote an entire section of this book to this topic.

An enterprise framework for People Analytics

The goal of People Analytics is to learn so that improvements can be made and investments protected. Uniting organisational objectives with how people do their work into a framework allows People Analytics to maximise and easily explain its value.

The range of potential applications for People Analytics is bewildering. That's not a bad thing, but how all the parts hang together can get confusing. Questions about which measures to focus on, how sophisticated you need to be, where to start, and how to show value can cause hesitation. I think what is needed in order to get a good handle on things is a basic conceptual model. Not quite an architecture that includes technology and data sources, but a simple and practical framework that brings together the different layers of People Analytics with a specific purpose - to demonstrably create value by applying insight into how people work and how the organisation behaves (see Figure 1).

Business Outcomes

The mission of the organisation is a natural starting point, and translating it into one or more performance goals and measures is usually straightforward at a high level. These measures are usually inescapable as the executive will be asked about them all the time, and include things like profit, revenue, membership. social good or others. If you own the business, you can tweak these with a bit of freedom, but if not, the choices are not so varied.

Behavior

How the investment in time, talent and energy is managed is the major determiner of whether an organisation succeeds (Garton and Mankins). Given that it is often the largest cost of a business, maximizing how people work is the number one management challenge. Business outcomes are achieved when behavior supports them. By behavior I don't mean just the 'touchy-feely' interpersonal etiquette, but **how** the work that is done, in its broad sense. If management is trying to achieve anything, it should be directed at influencing how work is tackled, *i.e.* behavior.

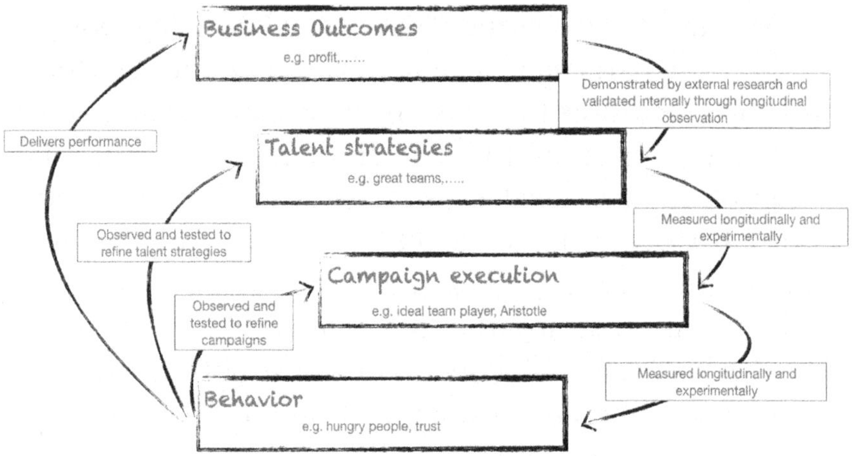

Figure 1: A People Analytics framework for influencing collective behavior to achieve business outcomes.

However, there is a wide gulf between high level organisational outcomes, and workforce behavior and this is bridged by talent strategies and campaign execution.

Talent strategies

The first layer connecting behavior with business outcomes is in defining a set of talent strategies or priorities. Just as executives plan how to accomplish the mission, there will be a range of decisions that will need to be made about the workforce and what it should do, and not all things can be prioritised. Is it the priority to source top tier talent over building excellent teamwork, is it to equip people with bleeding edge technology support, or is it to standardise roles? Is it to invest in leadership or to build methods to support processes? It can't all be done and so determining the talent priorities is critical, and may be the difference between business success or failure.

These strategies will be an expression of the intended business outcomes, but also start shaping what management activities will be done to shape behavior. Initially, there may be insufficient data to help define these strategies, but careful analysis over time can reveal whether the selected strategies are helping or hindering.

Campaign execution

The next layer is the implementation of the talent strategies as a set of campaigns or talent initiatives designed to influence workplace behavior. They may be for example involve:

- building awareness and adoption of new policies and practices relating to pay equity,
- training and measurement of agile team practices
- launching a new statement of leadership behavioral standards and expectations
- The inclusion of more opportunities for less formal, fun interaction or regular physical catch-ups

- greater development opportunity with additional resources made available
- continuously 'onboarding' all staff with light touch methods of maintaining organisational awareness

The key to campaigns is that they are intentional and implementable, and designed to influence behavior in a way that can be measured. They may be short or long term, focussed on individuals, groups or sections of the organisation. It may not matter whether they be physical or virtual, as long as they express the talent strategy, and seek to influence behavior these are the tangible things that are done with people to achieve business outcomes.

The links between layers

You may have noticed that this framework is a hierarchy, with strategy cascading down one layer at a time down to influence behavior. However, it isn't symmetrical with behaviour cascading up to strategy. Rather it is linked directly to all other layers. It determines business outcomes, but also informs the talent strategy and campaign layers. Data gleaned about behaviour helps determine whether the current strategies and campaigns are working and cost-effective, and whether they should be continued, changed or dropped.

Measuring the linkages between these layers is the task and challenge for People Analytics teams in enterprises. Incorporating external research with in-house experimentation, longitudinally attributing business performance to the implementation of a particular talent strategy, measuring behavior accurately and with as little disruption to work as possible as all challenges that need to be tackled.

Rather than conducting a range of analysis tasks that are distinct and disconnected, by continuously linking corporate objective with individual and team behaviours via talent strategies and campaigns will help People Analytics to continue to create and communicate value. There are likely many other frameworks that are more prescriptive, but anything that harnesses the investment in People Analytics toward organisational goals rather than just building out a set of analysis tasks and reports will help businesses decide where to start and what to do next .

- **Are you able to link people's behavior to business outcomes empirically, or does it happen purely on an intuitive level?**
- **Are you using data to adapt your management practices?**
- **Do you measure the impact of 'people' initiatives on business outcomes?**

AGILE HUMAN RESOURCE MANAGEMENT

"Best practice" defines those things that are observed to work well across businesses considered as leaders. It is a a useful approximation of practices that will work in other organisations. However, "one-size" does not fit all situations. Being data-driven and conducting experiments give businesses a chance to find out what works for them first-hand. Powered by analytics, and conversational AI, the task of learning bespoke lessons is far simpler than it used to be and fresh insight can help road-test ideas and iterate on them to create "best-fit" solutions.

Water bears and viruses

Those businesses that acknowledge the inevitability of change and build mechanisms to learn and adapt are those that survive. In light of the insight that data-driven techniques such as AI and People Analytics can provide, businesses also need flexible concepts and explicit patterns of adaptation that can harness these lessons.

Biological evolution is a game of winners and losers, a contest for the survival of the fittest. There are two ways of playing the game it turns out. The first way is to insulate oneself from change. Doing so requires a pattern of existence and reproduction that works over a very long time. It requires being resilient to change, even to future unknown conditions. That is, it requires splendid fortune or perfect foresight. The second approach is to breed quickly, experience some failure, and evolve fast. If this is your way as a species, you can't hold too tightly to any specific pattern or practice because odds are, in the medium to long run, not letting go will make you fail.

The Tardigrade is perhaps the animal I find most fascinating. These invertebrates can live in hot sulfurous volcanic conditions, deep ocean fissures under tremendous pressure, in moist and temperate mosses, and amazingly in the vacuum and the high radiation load of space. If you have not seen microscopic images of them, you really should have a look immediately. Photos reveal them to be extremely cute (hence the common names of Water Bears or Moss Piglets), resembling a character from an early "Doctor Who" episode. These guys do not need to change - whatever the situation is, they can ride it out. They are 'extremophiles,' which in some ways are the invertebrate equivalent of the ultimate multi-national blue-chip business that supports the basics of life. Just try and stop them!

On the other hand, are the viruses. Sadly, you probably know more about them than you did before 2020. Despite their ubiquity (being found everywhere life exists), and their abundance (one teaspoon of seawater contains approximately 100 million viruses), they remain at some level quite mysterious. That they don't quite fit into any of the definitions of like that are not concocted to explicitly include them is a clue to just how odd they are, and theories of their origin are varied. While they have encoded nucleotide sequences (as RNA or DNA), they cannot reproduce without hijacking the cellular machinery of a host organism. And yet when they do, they are prone to high degrees of mutation. When they find a niche that they can succeed in, they scale fast and spread. Viruses are the organic equivalent of the tech startups of the microbial world. Most of the variations fail fast, but when they get the formula right, they disrupt.

The business environment in which we find ourselves is going through the equivalent of a global environmental cataclysm. In the same way, we talk about the Jurassic and the Cretaceous periods; we are already using pre-COVID and post-COVID as shorthand terms for different eras. To operate successfully on the other side of this mass extinction event, businesses will need to either resiliently ride it out or to "embrace perpetual work reinvention" (Jesuthasan *et al.*, 2020).

- **Does your business understand its vulnerabilities and weaknesses?**
- **Are you evolving as an organisation, and if so, are you getting more 'fit' for what you expect the future will hold?**
- **Do you have any apparatus to discover whether your transformation is making you more fit or less fit?**
- **Is there anything that can be improved to help increase the velocity of adaptation?**

Every moment is an experiment

A business is a laboratory. Everything that happens has valuable lessons for the future. Few businesses have engineered themselves to be able to learn these lessons quickly and efficiently.

I love cooking. I don't get to do it often, and usually, when I do, I am mostly using left-over ingredients. It is satisfying to figure out how to blend them into a meal that my family will enjoy, and on rare occasions, that is what happens. For me, the kitchen is a kind of laboratory where I can explore what works. Unfortunately I never record what I do. Everything is a one-off. This is a parable for the workplace. What most people see in the office is work happening. What I see are hundreds of tiny experiments, and just like my efforts in the kitchen, the results of these experiments are rarely recorded, or lessons learnt from them.

In the workplace we tend to hire managers because of their experience and knowledge. Over the years they have witnessed a few things first hand and have stored some of the lessons away. This knowledge, and associated habits are from time-to-time at work and helpful things can happen. However, every manager is different, and as we have seen most are often flawed. Perhaps this is because in reality they have had relatively few experiences to draw upon, even after a whole lifetime, and no one has all the knowledge, especially as the requirements of work keeps changing - the shift to work-from-home is a good example of that. The knowledge of how to do tasks hasn't changed overnight but many of the elements of looking after teams has. The management hacks that used to work no longer do. What is needed is on the job learning through observation, and experimentation - however this trait rarely on display from an often overwhelmed management tier.

Research is not the exclusive realm of academia, and universities long for access to 'real-world data' against which to develop and test their hypotheses. Even if you are a small business, your workplace would generate more than enough data to keep your regular academic busy for an entire career. Viva la citizen scientist!

Jeff Bezos knew this when he set up Amazon. Whatever you think about his work, he learned from his algorithmic trading days that you could use data to experiment and optimise a business with - any business. There were plenty of reasons he started with books, but the intent was always to build a platform that could support a whole range of commercial activities. This enterprise included making the platform itself commercially available to others for them to build their own businesses on. It is called Amazon Web Services. The platform is a digital commerce platform, with measurement baked-in from top to bottom. Every transaction (including incomplete purchases) that it performs is a micro-experiment. As more transactions are processed, more data is available for lessons to be learned. Different advertising words, placements, pricing plans, and product features could be experimented with, to drive better business results. That things weren't perfect at the start was no cause for concern. What did matter was launching early, not just to start the money flowing, but so that the experiments could begin and the business could continue to be improved. The same principles can also apply to the workplace, where the data generated by its operations can help improve how it functions.

It's strange to me, given how long we spend in the workplace, and how ‘high-stakes’ work is for most of our sense of significance, just how little organised effort has gone into evaluating the performance of the people management practices we use. In cooking, sport, finance, health, entertainment, the effectiveness of just about every endeavor in life are subjected to analysis in one form or another. However, businesses have appeared resistant to being empirical when it comes to how they work, and have instead shown a high degree of inertia, clinging to standard methods of performance management, team structures, communication practices, and hiring. What we do today is essentially what we did almost 30 years ago (Fried and Hansson, *Rework*). It's as though somehow there was is mass conspiracy against testing alternatives. The rapid shift that were are in the midst of is seeing leading businesses apply data-driven approaches to designing *how* they work, and brought the effectiveness of work practices into greater focus. This new era is the birth of the workplace laboratory.

With data comes an opportunity to learn. We no longer need, and potentially are better off without, the expert-manager, and experiments are much more do-able now that we have methods of measurement at scale. It means that you can learn quickly, test hypothesis, abandon what doesn’t work, and double-down on things that do work.

In 'The Social Organisation” Jon Ingham describes how we need to move from best-practice to best-fit. It's important to learn from others because it's efficient to take their lessons and avoid the same mistakes. But what works in one place doesn't always transfer elsewhere. Until the lessons have been validated in your organisation, they are hypotheses whose worth is yet to be proven in your unique circumstances, no matter how meticulous the research behind them is. It makes sense to adopt them cautiously and test them as you do. No matter how small or unsophisticated your experiment, collecting data before rolling it out at scale is a data-driven approach that delivers 'best-fit' rather than 'best-practice.'

As the rate of workplace change speeds up, high velocity experimentation will become a core competitive capability, and this will involve learning at speed from others and from your own data. Currently, organisations excel when it comes to learning lessons from others, usually distilled by consultancies. The evidence for this is the proliferation of 'best-practice' guides on every management topic you can think of. However, organisations are rather weak when it comes to learning from experience, and rarely have in-house capabilities to perform this work. Establishing these capabilities comes with cost, but considering a businesses investment in talent, does this not make sense to do? I'm not suggesting that businesses perform research at the scale of a university, but maybe its time to dedicate some resources toward maximising the return on the 70% of your operating costs that is talent? A tiny improvement will pay for itself quickly.

- **Does how well your organisation is working matter and if so does your business invest enough in understanding this?**
- **Is there anything that you are planning on rolling out soon that you could do a small, quick or inexpensive test to validate first?**
- **How often does the appeal to 'best-practice', or 'personal experience' win an argument in your organisation? How often do people appeal to data derived from your own organisations experience?**

A culture of experimentation

Establishing a culture in which experimentation is normal is hard, and often, a significant overhaul in corporate thinking is required. As an analyst, one of the most disappointing experiences I have had is showing what I believed was irresistible and valuable insight, but for it to slowly compost away in a virtual pile of ideas. Putting lessons to work requires an organisational willingness and expectation that the organisation will act based on data rather than acting on the basis of inertia or opinion and this expectation is driven by a cultural habit of continuous improvement. As I see it, there are two critical challenges involved in experimentation becoming the norm.

- Firstly, most managers believe that how they go about managing talent is their business, and is an inalienable right and expression of who they are. They might see attempts to bring about change as misguided interference, and it is hard to argue with that perspective in the absence of data. While there will be many aspects of management and leadership that are stylistic and ought to be preserved, until there is an acknowledgment that practices can and should be scrutinised and improved upon, it will be an uphill battle to implement a culture of learning.
- Secondly, and sometimes as a consequence of the first issue, is the perception that when it comes to management, we just simply shouldn't experiment or 'muck about' with it. Random acts of management can get in the way, and adopting the latest fad can erode any willingness of the workforce to come along for the 'journey', but well considered experiments that create data to justify further adoption (or abandonment) of practices is a different story. Hesitancy can lead to stagnancy and becoming trapped with existing practices despite any evidence of their utility. Perhaps this is why digital transformations have most often focused initially on customer service processes and technology rather than on supporting the employee experience?

- **Do new ideas for how your people work get suggested, and tested?**
- **Do managers in your organisation consider how effective they are in their work? Are they continuously improving? How do they know?**

Agile learning

Being agile is about continuously learning through doing. Businesses have invested substantial effort and cost on doing 'agile' rather than on developing agile practices. Focussing on building learning habits and establishing processes for extracting and applying knowledge from experience are rare business priorities, but by prioritising these capabilities businesses can evolve positively, intentionally and sustainably - critical qualities in volatile times.

If you have been paying just a little bit of attention, you will have seen capital "A" Agile programs all over the place. The Agile movement (described in the Agile Manifesto) sketches out a sensible way of building software. At inception in 2001, the principles it embodied flew in the face of traditional project management, and pleasingly many of those lessons have been, or are at least starting to make their way into other business domains, including human resource management. There are many competing (and incorrect) definitions of agile that focus on speed, and a lack of disciplined planning, but at its core, "Agile" requires being open-minded and pragmatic about what works, rather than ideologically motivated. It is about getting real-world data to validate or invalidate assumptions before wasting too much time and money, and emphasises the use of learning cycles rather than highly detailed long-terms plans in the face of the unknown. The goal is to end up with solving genuine problems sooner, rather than addressing perceived issues that don't genuinely matter, later. Often, we hear Agile bracketed with Lean (Ries) as they share many elements to find productive solutions to problems.

As a former scientist, I can see that Agile has a lot in common with the scientific method, as it is ultimately data-driven and requires that ideas (projects) are abandoned when the facts change. What could be more suitable for adapting our talent management practices? In “HR from the Outside in” (Ulrich), a major line of argument is that HR needs to focus on the delivering value to the business, rather than executing a set of pre-defined functions, processes and responsibilities. The focus, should be as I have outlined in Figure 1 on solving organisational problems that are grounded in talent, not solving HR issues. Ingham (2017) also argues that new models of work need to reflect individual organisations' actual needs that are specific to their own work styles, rather than addressing imagined or assumed ones. These thought leaders point toward the need to be data-driven, experiment-oriented and adaptive. Put another way, they encourage human resource management to be agile.

Reducing the blast-zone

Rightly, business leaders get anxious with the idea that the organisation that they are accountable for becomes the subject of experimentation. “Isn't that too risky?” The last thing anyone wants to do is expose the entire enterprise to increased risk. “Business is hard enough already!”

This is a practical dilemma. Staying the same presents risk. Conducting experiments presents risk. The challenge is therefore in designing an approach to help strike a balance.

Ideally, experiments provide quick lessons without causing too much disruption. The best place for this to happen initially is in people's minds, where it is the safest to do so and the grey matter can trouble-shoot. Unfortunately 'in yourhead' reflects what is already understood and doesn't generate data. Stepping into the real world with experimentation will naturally create some surprises (otherwise why do it) so business leaders who think that surprise unacceptable should probably avoid experiments. Instead they rely on always being right about everything in advance. But for those willing to take on some risk, consideration should be given to how to minimise the potential 'blast zone' of tests.

Split testing

Split testing (or A/B testing) is a widespread technique used in marketing, as it is a simple and inexpensive way of testing the attractiveness of new offers or messages to the market. It involves splitting a target audience into two or more groups (hence A/B), making different offers to them, and using a predetermined metric to see which worked the best. For example, do people open the unsolicited email with the subject line 'Save the planet this winter,' or 'Save $100s on your electricity bill'. Finding out is different from guessing, and it is a simple matter of sending the messages to the audiences, waiting and measuring the differences between A and B. Did they ignore or open it, did they click the embedded link, and did they sign up for the offer? Thankfully that's a simple task.

For people management practices, this approach is rarely used, except occasionally in higher volume/lower risk transactional processes. In recruitment, modifying job titles and advertising, testing different work conditions, and changing where you cast your net can all be done relatively quickly. Businesses sometimes also optimise their internal communications this way, with invitations to awareness events or training sessions as well as company announcements having variants, to test which messages best "cut-through". However, the the same thinking could be applied to redesigning processes, policies, reward programs, staff development and a multitude of other applications. If its easy to run alternatives in parallel and measure impact - then why not?

Champion Challenger

A related experimental design is the champion-challenger model. Like split testing, different groups are used. Where it differs is that an incumbent "champion" approach continues for most of the audience, while a "challenger" approach is introduced little by little. Depending on which wins, you may stick with your champion, or the challenger may become the new champion. and await the day when it is itself challenged.

Early Adopters

In every organisation there are some early adopters; those people who are always keen for change and like to get involved in anything new. Typically, this group is easy to ring-fence and test new ideas with, introducing them to something new and measuring feedback. It's tempting to use them because it is relatively pain-free, and it minimises disruption and complaints from the laggards. The strength in the approach is that the open-mindedness to change means that you can quickly gain insight into a potential future state. If early adopters can't see it working, it probably wont. However, the weakness is that these group isn't representative of the whole organisation and they might provide unrealistic, "rose-colored" perspectives and painful lessons will only be learnt late in the process.

Natural experiments

Organisations are unwittingly or intentionally experimenting with their talent all the time. For example people occasionally find themselves working in different teams with different styles or managers. Groups of people are located in separate offices and locations, or find themselves working on different tasks, projects, or shifts. Employees also put themselves into groups such as those who prefer to work earlier than others, and those who select one technology over another. The list goes on. All of these groups are examples of where natural experiments can take place, but rarely are differences in important metrics measured. Sometimes you have the luxury of being able to set up nicely randomised double-blind or blocked control groups, but mostly you don't, and so using natural groups is a very practical approach. By intentionally measuring business-relevant metrics associated with whatever groups you've got, it's possible to learn which practices are a better-fit.

Hearing the Agile Jester

Having spent many years in the tech sector, but with a continuous involvement in Talent Management, I'm interested to see that #agileHR has gained much traction on social platforms.

While the principles of agile are simple, some organisations have struggled to adopt them. Rather, mechanistic methods, rules and processes have been developed to try help implement the principles. Unfortunately, these have often obscured the essential ingredients, resulting in an "Agile" that is in name only but lacks the main principles. My favorite characterization of this kind of window-dressing is that of the Cargo Cults (Kejzlar), who tried replicating some of the manifestations of post-war recovery (including radio communication, planes and runways) using the low-tech, jungle materials they had at hand. Just because a coconut is fashioned into a radio doesn't mean cargo is on its way. It seems like a whimsical parallel until you live in agile hell (Martin). I think that every organisation that adopts a formal agile framework needs to provide someone with the truth-telling mantle of the "Agile Jester". This would be a person that is allowed and expected to step outside the jargon, and to make light of it when it goes wrong - for the purposes of correction. The Agile Jester can point out when the emperor has no clothes, and keep the efforts to be Agile on track.

The software industry is now operating in a 'post-agile' world, with some of the original proponents now admitting an agile defeat of sorts (Jeffries); not on the basis of failed principles, but based on the term becoming so polluted that it's hard to discern what it was really all about in the first place. Nevertheless, there is a deluge of blogs describing the fact that HR is 'going' agile. A quick scan of YouTube for Agile HR will offer a playlist of case studies about new structures and functions, but very little about testing new services with users, getting behind people's perceived needs, or how a series of incremental experiments has led to unexpected approaches to freshly uncovered problems. I fear that as with software, with #AgileHR we are witnessing a structural and procedural change rather than a data-driven one. If this is the case it will probably peter out quickly, but my hope is that the growth in #AgileHR will see data being harnessed with processes to support rapid learning.

- **Are people in your organisation suggesting that you need to be more agile?**
- **If so, are they using agile as an adjective for how you should operate, or as a noun, signify a framework they want you to invest in?**

Continuous Improvement and the Captain's log

If mistakes are not to be repeated, it's necessary to keep a record of them. If lessons are to be shared, then records need to be interpretable by others. This means as with a scientific journal, the conditions at the time of an experiment also need to be defined so that experiments can be repeated and challenged if necessary, or when conditions change. Business is not academia, and competing priorities will probably inhibit the rigour of such records, but they needn't be complex to be useful. Marketeers are fortunate because the platforms they use usually have the necessary analytics to record the effectiveness of campaigns that they have run, but HR practitioners rarely have access to data that so obviously flows onto business goals. For those responsible for talent strategies, the measures that they have are not standard, and those that are common don't help. Test-and-learn is tricky, and keeping all of the data organised about the hypotheses, interventions, treatments, controls, the results, their timing, and their interpretation can be complicated. Despite this, practitioners need to start somewhere and a very basic framework can help organise these activities for those that are yet to start.

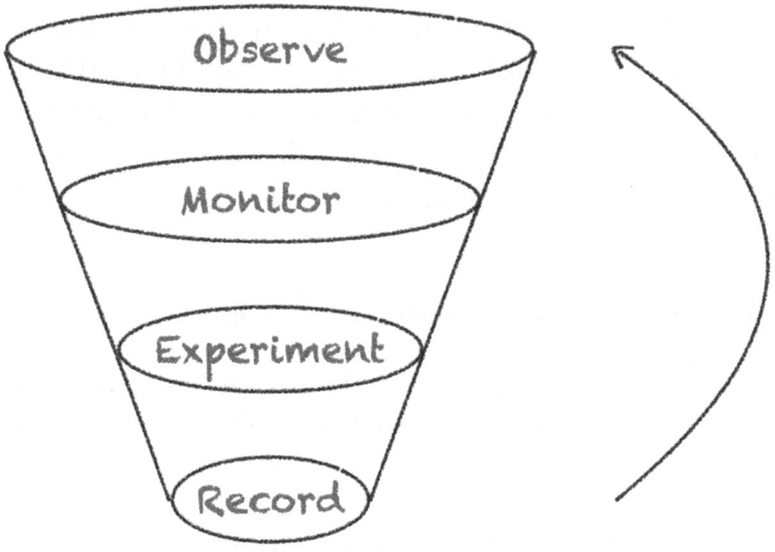

Figure 2: The workplace laboratory funnel

Figure two outlines a process of continuous improvement that culminates in the creation of a record of lessons learnt. It starts at the top of the funnel at ***'Observe'.*** This is where curiosity plays a role, leading to the formulation of hypotheses. Observations might arise incidentally, or through data-driven monitoring and exploration. Observation is usually accompanied by statements such as 'I wonder if.....'.

For those observations that are of most interest or value, a process to ***'Monitor'*** the data can be set up, without jumping headlong into to the trouble and expense of formal experimentation. This provides a chance to work out how the idea can be measured (and if it can't then maybe leave it as an interesting idea) and to simply how the metrics change over time and amongst groups. This might be sufficient in resolving any uncertainties and provide sufficient answer to the initial questions posed. For those questions that remain unsolved a more formal ***'Experiment'*** stage might be justified. Along with having already established metrics that matter, and possibly some expertise with tracking those measures, the process of conducting an experiment might be greatly simplified.

This framework is simple and is designed to help businesses that want to start being intentional about what they measure. It provides a basic process and a set of stages to help manage the investment required to become data-driven and might be especially useful when hitching HR to the "agile" bandwagon.

- **How does you business record lessons learned, if at all?**
- **Is there a body of knowledge that you use, and if so, do you contribute to it?**
- **When considering change, does your organisation assess it from what it has already tried?**

Some worthwhile experiments

Experiments are easy to do with the right mindset. They needn't be complicated or require a PhD in science to perform. Experimentation is about being intentional about continuous improvement, and accelerating the rate of learning by testing ideas in the real world and within specific conditions.

Here are some experiments for you to think about with your organisations. They may not work in every situation, but one or two may be relevant - hopefully they trigger a number of better ideas for you to pursue.

- Run two job advertisements for the same role in parallel, once with conservative language, and one that is more playful. Do it before you actually want to hire someone and don't spend any cash on it. See what kind of responses you receive, and try it for different types of roles. Rate the responses into simple 'good,' 'OK,' 'bad' categories, and work out which approach worked best. Now when you advertise for real, you can do this based on what you have learned, confident that some of your hiring practices have been improved.
- Run a long update meeting full of intricate but relevant detail, avoiding the management spin. Later run a

short session without any detail, but just what you think the top level messages are. After each meeting, collect feedback on whether the meeting was useful, helped people understand, whether they found it motivating, or enjoyed it. Share the results with everyone and ask for suggestions about the next meeting and when it should be.

- Perform a survey and ask people to nominate whether they feel isolated or not. Now count how many days per week each of the respondents works remotely and categorise them as 'frequently remote' and 'frequently in the office' to compare feelings of isolation. Next randomly assign all individuals into two groups. The first group will receive an unscheduled phone call from a nominated person (their manager, manager's manager, CEO, or someone from HR) and the second group left alone. Rerun the survey one week after the phone call and see whether the sense of isolation has changed. You might consider a different intervention; like sending flowers or inviting them to an informal gathering. If things improve, bake it into standard management practices.
- Take a chat application and count how many rooms/groups/teams/channels/contacts each person is a member of. Keep the result handy. Often people believe that collaboration matters and is enhanced by being connected online. Run an event where different teams showcase their work to one-another and later re-run the analysis. Did connected-ness on networks increase?
- Naturally, you will have people that respond to emails very quickly. Find them by getting the email metadata and split them into fast responders (group A) and the rest (group B). From those groups, calculate the average performance ratings last time they were measured (you might need to break the set into job families first). Are those that respond quickly your top performers or the other way around? Maybe do this just for the managers. Explain the

results in general terms to everyone and ask why they think it was the case. Afterwards see the impact on responsiveness.

- Run a short multi-choice test about an important policy that has been published and measure awareness and understanding. Let a policy-bot learn the policy so that it can respond to employee queries. Get the policy-bot to send occasional reminders to staff about key details, to gauge feedback about it, and to test understanding. After a while re-run the multi-choice test and see whether there has been an overall improvement. What's the return on investment for this activity (10% improvement in understanding for $10 per person?) Is it worth continuing or expanding to other policies?
- Set up an awareness-bot that everyone can train with their top FAQs. Monitor usage and track satisfaction with responses. Let people see the data so that they can keep training the bot. Monitor usage.
- Before using the entire training budget on one program, send a few people to a couple of alternatives. Ask them about it and get them to decide. If you want to be more objective, measure feedback from colleagues about the behaviour you are training for continuously throughout the programs. Determine which has the greatest bang for buck and roll it out more broadly while measuring the impact more broadly.
- Execute a campaign measuring team effectiveness. Separate the teams based on size, distinguishing between large and small groups. See whether there is a peak team size based on effectiveness. Now in the future, consider how large teams might be structured into smaller units and vice versa. Continue measuring performance with the same campaign and validate whether tinkering with team size has been useful. If not, cease worrying about it.

- Enforce team member 'time out' while on client assignment, and make them unavailable for set periods. Do it by introducing this change to some teams but leaving others as control groups. Measure the quality of work via simple longitudinal check-ins and surveys. Sounds like a disaster, but BCG did this and found that service quality improved, and led to better work-life balance for staff (Perlow and Porter). Boston Consulting Group did this explicitly within an experimental framework and tested a number of other hypotheses as part of a process for continuous improvement.

Experiments don't have to be hard, and measurement does not need to be sophisticated to be useful. Small experiments can become part of the background activity of a business without causing major disruption or workplace distraction. Why not include the results of them in your monthly updates to build a culture of being data-driven?

- **When was the last time you stopped a management practice because you had evidence it wasn't working?**
- **Does your organisation actively consider experiments that it can conduct to learn better methods?**
- **What about your workplace is the most surprising productivity enhancer that you have discovered?**

A growth hacking mindset

Growth hacking is an approach to engaging with a community (of potential customers). By applying the same thinking within the workplace, it is possible to identify low-cost methods of engaging effectively with a large number of employees.

In 2010 the term "Growth Hacker" was coined to describe a person that focused on innovating low-cost but effective ways of reaching a large market with their products ("Growth Hacking", Aufray 2016). Until then, businesses would create a product, design a market launch, deliver the product, and look after customers, and eventually find out if they were successful. Sophisticated and well-funded businesses would include focus groups, early-adopters, and other market research techniques to guide the journey. Defining the target market early in the process was necessary because it was the only way to manage the risk that expensive marketing dollars would be wasted by launching something that they wouldn't buy. There was no second chance.

In contrast, when the growth hacker came along, they worried themselves less with advertising and more with understanding potential customers. They focused more on building a relationship through social channels and less on creating a recognizable brand through paid channels. They emphasised discovering which aspects of their product were a good fit for which market niche rather than on measuring a net promoter score. They were inclined to seek viral means of influencing consumer behavior more so than they were building a linear funnel of leads and prospects. They wore T-shirts rather than suits, and they knew tech as well as marketing.

If your workplace is like most, then chances are it is replete with the traditional people management practices that the growth hacker would roll their eyes at. It is worth questioning how you might currently be taking these practices 'to market' and to reach your 'audience' of employees, especially in the context of the role of managers. Later in this book I detail how some practices might look where armed with data, but as a thought experiment, what would you expect the growth hacker might think of the following ways of ‘reaching’ employees:

- Monthly all-hands meetings where the leadership outlines the results and reinforces the vision and strategy
- Intranet blogs announcing changes to the policies and procedures
- One-on-one discussions to retain key staff as their contracts approach their use-by date
- Weekly one-one-one manager discussions with staff to update progress and adjust activity
- Annual performance reviews and development planning with managers
- Filling out spreadsheets to update on progress and describe upcoming tasks

These are deeply entrenched practices in many businesses, but rarely is the usefulness quantified. Most of these activities seem to have some obvious merit on the surface, but in the broader scheme of work, people and managers in particular are busy doing other 'more important' tasks. If these activities are perceived as mandatory but annoying distractions, then they are doomed to fail as they will probably be done poorly. The growth hacker wouldn't respond to this situation by just trying harder, doubling-down and re-committing the management to 'making' them work. Instead, they would rethink the end-goals, and consider how else to achieve them. Perhaps they would by-pass the barriers to employees getting a proper development plan sorted with their manager, and test alternatives that are more direct.

For example, how did the growth hacker reach, lead, and educate a distracted and anonymous market? They built a relationship through social media, with careful, bite-sized content that consumers could nibble at, and request more if and when they wanted. It was by using lower cost, just-in-time and memorable messaging more than the beautifully printed banners. The great lesson was that unless potential consumers were already in the zone, 'interruption marketing' made them switch off and filter the distractions out (e.g. Hirsen, 2019). It is identical at work. In the office, it is a fight to get people to pay attention to announcements. It doesn't really matter if they get delivered in person, through email blasts, or via blogs, if you measure engagement with content you will probably see that it is fairly low. Connecting with and engaging your workforce was hard enough when they were in the office. Now that Netflix is on in the background, employees work when they feel like it, and can tune out without being noticed. Getting attention is tricky indeed, and managers, now physically dislocated from staff can't rely on proximity to exert control.

Creating content can also be expensive when you consider the direct costs and the effort involved. And if "cut-through" is achieved, then it can become even more costly to follow-up on the feedback that it triggers. These issues mean that we tend to do things infrequently, and only at times that we think are the most critical (and schedule-able) touch-points with employees. It is no wonder that the important is often sacrificed on the altar of urgency.

You may rightly protest that the workplace is not some marketing venture, and employees are not customers. It is a different relationship with reciprocal obligations, not the least of which is complying with labor and contract law, but do workers actually see it this way? Organisations rely on managers to deliver a fair and uniform experience, which has led to the development of processes and templates to make sure that the right things get done. Unfortunately, these templates have often become so detailed, so tailored to cater for every possible corner-case, so one-size-fits-all, and so little time is available that managers and their teams just pay lip-service to them. Mechanistically they endure the mandatory processes which they decipher as a set of check-boxes. While such templates achieve a level of paper compliance, they deliver little other actual business value. In this situation, the the goal of creating consistency means that the experience is consistently bad.

Dollar-for-dollar the growth hacker is able to reach and influence a much larger audience through social channels than traditional marketeers. Those advising on how the people aspects of work are operating (perhaps your Head of HR, People Operations, Chief People Officer, or line managers themselves) should adopt many of a growth hacker's qualities, knowing how to use the available tools to build relationships with a purpose.

- **Which channels do your employees notice? Which ones do you use to influence them? Is it working?**
- **Are your leader's announcements seen as a distraction by the workforce?**
- **What can your leadership do differently to influence your people in a low-cost way?**
- **Given a choice, how do your workers prefer to connect?**

Employee experience

By applying the techniques of "Human Centred Design" at work, organisations are being shaped to fulfil their mission in a manner that builds value for its employees. By bringing workers into every step of organisational design, the chance of creating systems, processes and structures that fail is reduced, and the noisy 'expert' opinion is virtually silenced. A renewed focus on employee experience is bringing financial and other benefits to business, but without data-driven underpinnings, these efforts will be unfocussed.

Customers are shifting their spend away from products and services to 'experiences'. Consumers now build relationships with suppliers rather than buy from them. Our minds tend to become programmed to associate experiences with brands, and we are happy to be loyal and pay a bit more of a premium for them. I suspect that this relates to moving up the hierarchy of needs from security and safety to self-actualization (Maslow, 1943) and investing in brands that help us tell ourselves the stories we want to hear about ourselves. It seems to me that the employee expectations of work are following the same trend - it's not enough to turn up and do your job in return for pay, it's more about finding somewhere that you can express your true-self in and feel good about who you are. The modern worker now expects to be able to use their unique talents to expertly handle their work, enter a state of flow, and find happiness. If the trend continues, then the meaningless rituals that get in the way won't be tolerated.

Mike Cooley is famous for kicking off a method for innovation termed human-centered systems (Cooley, 1989). Doing so was a reaction to the often lazy and autocratic approaches to design, and brought human reaction and response into every stage of the design process. It paved the way to Human-Centered-Design (HCD), advocated by Stanford's d.school, reflected now in a global standard (ISO 9241-210:2019). HCD is applied now to a wide array of design domains, including software design through to physical work environments. We have also seen a growing recognition of its utility from HR practitioners as employee experience (EX) programs get underway (Coolen and Sombezki, 2019).

Some might assume EX is a 'soft' management trend, but research by Young and Kulesa (2019) shows otherwise. In an analysis of five key financial metrics (return on assets, return on equity, 1-year change in gross profit margin, 3-year change in gross profit margin, and 3-year revenue growth), they observed a 5% improvement in each metric in those businesses with a strong EX program in comparison with those with poor EX. One of the drivers thought to be behind this was the improved Service Profit Chain (Heskett et al. 2008) which can be paraphrased as 'happy staff make happy customers'.

One of the critical requirements of EX is to inform design by including real human reactions to existing and potential situations. In this way, EX programs (and HCD more broadly) is a data-driven and experience-based stablemate of agile techniques, and has thusly boomed along with it. Its strength is the discipline and skill to genuinely get human input, which can be tempting to bypass under deadlines and dwindling budgets. The tendency to assume that people's reactions are well known already can jeopardise the design process, and while jumping into someone else's shoes seems like an obvious thing to do, kidding yourself that you have already done so is too easy. That HCD has grown to what it is today is proof of just how hard this task can be. This difficulty has particular relevance to re-inventing remote work where you can no longer 'bump' into people in corridors and over lunch. While the absence of a "water-cooler" makes little difference to the allocation and completion of tasks, the question about whether serendipity people think differently and consider broader issues is unclear. Either way, larger enterprises now often have a design, digital or EX team that is focussed on collecting and interpreting data directly from people, and often through digital means. This means that the days of relying on the manager cohort to garner sporadic feedback and to represent their people in design processes is unlikely to be necessary - or helpful.

A focus on the employee experience is good for workers and good for business, but without the necessary means to collect and interpret data, and to include that into continuous agile cycles of build, test and learn, employee experience programs will lack the ability to quantify their value. This will resign them to the same fate as other fads that come and go.

- **Does your business prioritise the employee experience when designing change**
- **Is employee experience part of your organisations' talent strategy**
- **Have methods for involving people in design and change programs been properly thought through, or is lip-service being paid instead?**

THE CONVERSATIONAL ENTERPRISE

The combined and multiplicative effects of data-driven approaches such as bots, People Analytics, and Agile HR spells the end of middle management as we know it.

Listen, Learn, Act

Broadly speaking, the act of managing an organisation is about listening, learning, and acting. The process of managing is about facilitating an ongoing dialogue. While the cost of communication grows with increased organisational size and complexity, a new data driven-approaches open up a scalable 'conversational' business architecture. I call this the "Conversational Enterprise".

The "Conversational Enterprise" is one that is data-driven at its core. It uses data to facilitate an ongoing dialogue in which it "listens" to the workplace using the techniques of People Analytics, it "learns" its way to successful operations by continually testing and learning, and it "acts" to drive change using Conversational AI.

Traditionally the Middle Managers played a vital role in each of these three elements. However, as outlined earlier, this approach is an expensive way to achieve business growth, leading to complexity and operational inconsistency. The role of Middle Managers has narrowed due to the adoption of new work practices and supporting technologies. Moreover, Middle Managers have complex 'human' responsibilities that they in general they have struggled to adequately perform, leading to high levels of staff disengagement and diminished trust in them.

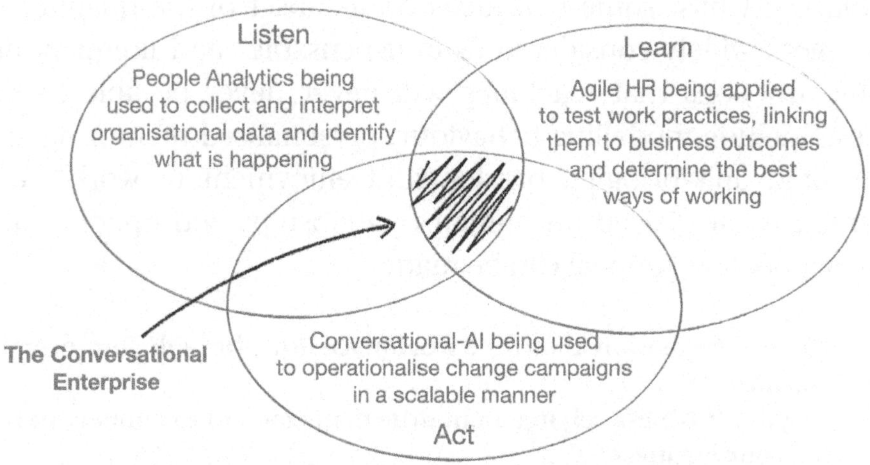

Figure 3: The conversational enterprise sits at the intersection of People Analytics, Agile HR and Conversational AI, and is a scalable method of organisational listening, learning and acting.

In the place of an architecture that relies on middle managers will emerge a new business architecture in which many of the functions currently considered 'management' are delegated to intelligent systems. Business growth will be less reliant on layers of middle-managers and more reliant on connecting teams through data, algorithms, and incremental cycles of change mediated by conversational bots.

People Analytics can now be used to identify important organizational dynamics and to distil employee voice. Agile HR can build a positive employee experience through careful experimentation, and Conversational AI can provide connectivity between teams, individuals, systems, and processes. By combining these three fields, a business can help support the 'management' layer without adding managers, and therefore it is a highly scalable approach.

New roles will be formally identified to work within the Conversational Enterprise. Some of these will look after three capabilities, whilst some will focus on a subset of the qualities of managers widely considered as indispensable and uniquely human; functions that machines will never never be able to do. These include modelling behaviour, personalised mentoring and coaching, and bringing passion and enjoyment to work - elements that are based on human relationships and operate only when a personal appeal can be made.

- **Does information about the workplace flow through your organisation**
- **Is your business relying on middle management to connect parts of your business?**

Displacement or augmentation

Over time, various attempts to reduce the number of managers have been made - with mixed results. Due to changed work patterns and the 'Uber-ization' of work, the impediments are no longer as strong. Over the next decade, the middle manager layer will start to be replaced and augmented by AI.

Google once experimented with getting rid of the manager cohort. Seen by their own staff as bottlenecks, they ran an experiment to determine whether their removal led to increased productivity.

It didn't.

What emerged instead was conflict, demonstrating the valuable but less tangible qualities that managers can bring to work. Great managers have a hugely positive impact (Beck and Harter), and yet they are rare, and niggling questions about the future of management persist (Hamel). Whether managers will be replaced by robo-bosses is an open question. However, there is little doubt that the process of management will change, potentially to the point where the role of "manager" is unrecognisable.

A more hands-on executive

AI reduces the need for humans to perform highly repetitive tasks that are characteristic of unskilled, lower-paid, and junior roles (Frey and Osborne). Less recognised is the fact that AI also has great potential for increasing executive engagement with the workforce. Just as Donald J. Trump engaged directly with his constituents via Twitter, business leaders will be able to use technology - not just to make announcement, but to steer their organisations in a more hands-on manner. Many executives have already experienced better engagement with staff during work-from-home conditions, with mechanisms such as 'fireside chats' delivered via live and recorded video. Staff often also appreciate this increased contact. AI applications will also mean that executives will be able to participate in different kinds of interactions within the organisation, communicating direction, reminding of critical outcomes, collecting data, motivating, coaching, and measuring compliance. As Lencioni (2020) described, the CEO is the 'chief reminding officer.' Currently considered too time-consuming, bots that are delegating to act on behalf of the executive will simplify this task. There is little need for a manager to hear from the executive and pass that on to teams when the executive can instruct a bot to do so directly on their behalf. Jesuthasan *et al.* (2020) identified from research with over 500 business and HR leaders that business leaders are already adopting "an ambidextrous mindset driving strategic business initiatives in the C-suite while continuing to deliver flawlessly on administrative and operational tasks." Bots will have a pivotal role in helping them do this, by giving them the necessary support to allow them to be continuously involved in the management conversation.

The threats to middle management come from both sides, being squeezed from above and below. With fewer people to look after and a more hands on executive, what remains? The top and bottom layers of an organisation are always required. A strategy must be set and operationally executed, but the generalist activities of coordinating, escalating, reporting, and consolidating are optional, and possibly better performed by technology. It can only be their distinctively human qualities that bots cannot replace that remain, but as we have seen - managers are struggling to bring those qualities to work, and bots informed with data are becoming more capable.

Bots augment rather than substitute

The power of machines to perform a range of tasks, including those once the purview of managers is increasing. In the years ahead, swathes of workers will be replaced by AI and businesses seeking competitive advantage will look for increasingly sophisticated AI to further reduce costs. However, for the remnant workforce, they will need some ongoing human contact. Humans have evolved socially, and much of the time available to us in life passes while we are 'at' work. People that work in isolation suffer significant health challenges, and high-bandwidth collaboration within teams is best for creative problem-solving.

While intelligent machines will grow in capability, their role will probably only ever be complementary rather than substitutionary of middle managers. Exclusive reliance on bots where human contact is needed will harm an organisation and its people. It may be that those businesses that best leverage the precious, valuable (and relatively expensive) skills of humans will outperform their counterparts.

Complimentary skills

Google's experiment showed that there are things that people can do that aren't always obvious until they gone, and they are probably things that a bot will never be able to do for the workforce that people can. It's no surprise that Gallup has discovered that great managers excel in a range of related qualities. Motivating employees, showing assertiveness and drive in the face of opposition, creating clarity with respect to accountabilities, and prioritizing productivity over politics (Beck and Harter) all require a personal appeal, the strength of which is grounded in the manager-employee relationship. The flip-side of the adage that 'people leave managers not companies' is that they stay for them too. These are clues as to what the qualities of the future "manager" will be.

However, some results show somewhat surprisingly that bots can also excel in some 'human' aspects too. Roughly two-thirds of the workforce suggests that they would trust an algorithmic manager more than their current manager, which might just reflect the current poor state of management skills (Percy). Bots excel at responding immediately, no matter how 'busy' they are. They can learn the detail very quickly and can remember details without ongoing reminders. They can act and explain consistently and impartially, irrespective of the morning they have had. They can listen at scale, taking inputs from everyone as they are unfatigued or restricted by the hours available in a day. Bots are also extremely capable of connecting people with solutions, allocating tasks, and keeping track of progress. Even though bots are typically considered weak in solving one-off problems that they haven't encountered before, they are now showing the early signs of Artificial General Intelligence (AGI) and an ability to solve novel problems with learned rules of thumb. As the number of tasks that bots can do to a high standard increases, the relative reliance on human managers will decrease. It's unlikely that managers will disappear altogether, but their roles will probably be very different.

- **Are the executive in your organisation too busy to engage with the front-line workers to the degree they'd like to?**
- **What is it that your managers do at work? Are they coordinating people, developing them, busy with their own tasks, analysing performance, recommending change?**
- **What are the distinctive qualities that your workplace expects of its managers? Are they fulfilling them?**
- **What would happen if your managers stopped coming to work?**

The new roles needed

With the changing nature of work, and the rise of the Conversational Enterprise, new roles will emerge for both humans and bots.

Some view bots purely as digital assistants to act at the behest of their humans, while others predict that robo-bosses will dominate workplaces. These are simplistic ends of a spectrum, but the reality is likely to be that bots will start to appear in a range of different, and more narrowly defined roles, working alongside their human counterparts, sometimes replacing them and sometimes altering their function. In this chapter I describe some of the more specialised functions that people and bots will perform, and the new roles that will be needed in a Conversational Enterprise.

HUMAN ROLES

The bot-wrangler

In an earlier chapter *"Botty Training"* I described how the training and building of bots is becoming increasingly simple; from deep technical data-science and software engineering expertise to generic teaching skills that we might use amongst our fellow human co-workers.. Nevertheless, much like how HR are involved in defining standards by which people work, behave, dress etc, a role to help govern how bots work; limiting what they have access to or permission to perform, how they look and behave, and how they complete more complex tasks will probably emerge.

The mentor-coach

Football, basketball, and elite team sports invest considerable sums in their talent. It's no wonder that they maximise the return by also investing in coaching staff. Teams have a primary coach, offence coach, defence coach, special skill coach, high-performance coach, dietician/s, psychologist/s, coaches assistant/s, and probably a coach's coach. No wonder being a coach is lucrative in these sports.

In the workplace, we also invest considerable sums in talent, but where are the coaches? Workplaces have managers and executives, but when you look at these roles, they often spend more time with one another than they do with their teams practicing, talking tactics and reviewing 'game' performance. Who focuses on getting the best out of people? The answer is often 'nobody'. Managers today appear to spend most of their time representing the team in decision-making, communicating status, organizing resource levels, and protecting funding - activities that happen outside of the group. Some businesses are already moving toward a state where managers are not experts but enablers (Ibarra and Scoular). But by and large, businesses lack coaches, and the opportunities to coach have become even more limited with remote work. The opportunities for the spontaneous moments that culminated in 'why don't you just swing by and we'll chat,' have dried up.

Coaches also play a role in keeping a lid on potential powder kegs and flame wars. They are usually selected based on their ability to keep superstars and their gargantuan egos from fighting one-another, and focused instead on playing well together on the field - where the results matter. Such hostilities and power imbalances can escalate quickly when everyone is mostly communicating by text. While team incentives and structures can help teams self-moderate, a coach may need to be present to manage a team charter, execute team rituals, and provide a steady hand when dealing with pressure. The experiment from Google referenced earlier ran aground because of the coaching deficit, and points to a need for more hands-on coaching within teams. Coaches also motivate and understand what it takes to get the best out of their roster, which may differ from player/worker to player/worker.

A portion of today's managers may become tomorrow's coaches. To coach more intently, these people will need to offload some of their current coordination responsibilities into intelligent applications. This might also allow them to support bigger groups of people.

BOT ROLES

The coach's assistant/development aide

Like the executive, the newly established role of workplace coaches may need to help in doing their work. Figure four provides a schematic of the kind of assistance that they could expect to receive from a coach's assistant or development aide. Having been instructed by the coach, they will be deployed into the regular communication channels as everyone else is using to interact in. They may explain a range of policy details, announce changes to the operating processes or arrangements, measure the impact of new practices, test knowledge and understanding, and also listen to and distill team feedback. They will run drill with the team members, highlighting areas for team and personal development, and reference materials and methods that can be used to support improvement. In what Bersin (2019) describes as 'employee engagement 3.0', we see outlined that the next wave of People Analytics will be focused on delivering such outputs in software applications. Bots in channels will be able to personalise their delivery to workers, with little distraction, as they work.

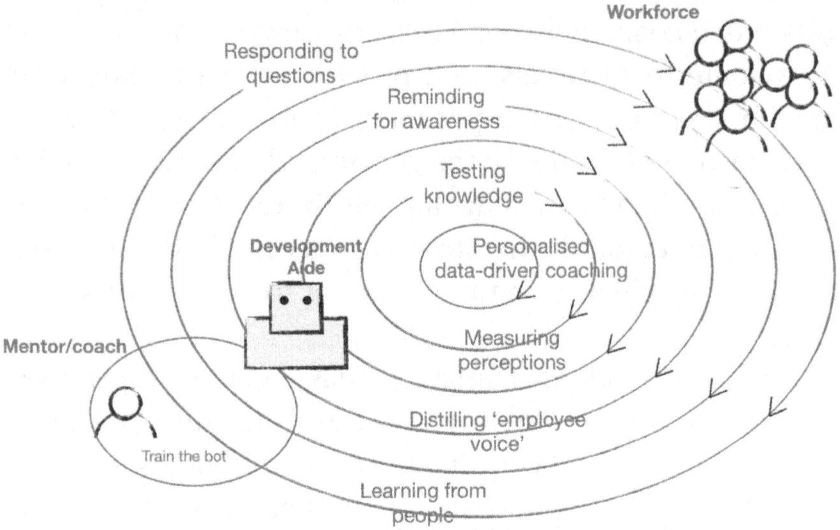

Figure 4: Schematic of a human coach and a development-aide at work

The coordinator-bot

A key strength of bots is understanding the detail, all of it, and staying up to date with it as it changes. By knowing how the various parts of organisations connect, bots will be able to support individuals with expert and detailed guidance about how to connect. Whether people need to understand roles and responsibilities, engage with processes and policies that are only occasionally relevant and so easily forgotten, or to track tasks and dependencies, coordinator bots will be able to steer people and build cohesion.

The personal assistant-bot

As our digital assistants already do in our phones and smart speakers, personal assistants will help workers do their jobs more fully in the years ahead. Responding to short precise commands, digital assistants can execute tasks and gather information for individuals to make them more productive in their work. And as the bots grow in sophistication, their creativeness, thoroughness, and speed will make them indispensable at work at many levels, allowing people to focus more by spending less time switching between tasks.

The relationship here is likely to be very strong, given the personal nature of the assistance provided. We will probably see these bots with highly customised personalities to cater to individual preferences.

The delegate-bot

As bots become as simple to train as their human counterparts, all employees will be able to make use of them so that they may stand-in and work on their behalf, representing them in conversations when people are away, or would rather focus on other things. The quality of their input will however be a function of how well they are taught. It would be nice to be able to focus, free of the distraction of responding to queries, but knowing that people are getting what they need.

Smart coworkers

As digital assistants evolve and as algorithms and datasets improve, bots will become increasingly proficient at niche tasks and operate independently, with new better-than-human skills. Bots are already proficient at collecting and analyzing data, working as coworkers taking the initiative with teams associated with them (Griffin and Coleman). Equipped with specialized sensors, bots will be able to monitor and either act, recommend or simply escalate their observations to the teams that they are part of. It will be like having an autopilot in a team able to advise on strategic and operational matters based on the situation. Every team will have unlimited access to bots that have deep expertise in technical areas.

These workers will either increase teams' capability by adding powerful virtual workers to them, or shrink the teams by replacing individual team members. This equation will depend on the organization but based on history, we shouldn't be surprised if they do both.

- **Does your business invest in coaching its people?**
- **If bots became more widespread in your workplace - who would look after them?**

The Org Chart of the Conversational Enterprise

Organisation structure increases in complexity with growth, as new layers, dotted-line reporting, and special arrangements are made. The introduction of AI and the evolution of human roles in the management layer, along with a more direct connection between the front-line and the executive provides an opportunity to greatly simplify structure, allowing a business to grow numerically without adding structural complexity.

The most obvious implication of the Conversational Enterprise on the structure of organisations is that the human-management layer will shrink, and disproportionately more so than the front-line. As it does so it will focus more on supporting the human needs of staff, such as motivation, and more complex and personal issues the way coaches do. Gartner predicts that this transition will be gradual, with algorithmic management working to increase the span of control of managers, who do not get replaced (Goasduff) except through natural attrition. I like to think of this process as replacing the span of control with a network of influence. Control will be less of a goal as teams manage themselves, and relationships will not be hierarchically arranged.

While the Conversational Enterprise requires fewer people overall, and businesses may see this as a virtue in itself, this architecture is designed to make use of the best of people and bots to create a simple, resilient and scalable approach to building alignment. Adopting it also happens to create the conditions that support workplace experimentation without disrupting business. Change can be difficult and slow as information cascades through and behavioural habits die hard. Bots can consistently reinforce and measure change. Rather than just having 'digitally dexterous' people to support change, the Conversational Enterprise is organisationally dexterous.

Figure five provides a simple version of a hypothetical org chart under four different scenarios (A-D) for comparison's sake. In scenario A (the traditional hierarchy), the organisation comprises several layers with individuals arranged in teams, reporting to managers. Managers may have multiple teams (rather than just being team leaders), and they act as a bridge between layers. They are the 'meat in the sandwich.' Each manager may have a different style and knowledge about the rest of the organisation. They play a crucial role in maintaining order via command and control mechanisms, and how connected teams are is a function of their managers' individual and collective abilities.

In scenario B the Conversational Enterprise, connectivity between individuals and the rest of the organisation (including the executive) is mediated by AI bots that coordinate and communicate between the executive and the frontline. I use the term augmented as it is not intended to replace direct communications, it's just that communications can be bolstered with AI. In this scenario, there is no command-and-control manager of teams. Instead a human 'coach' supports several self-managing teams by focussing on more generously supporting the needs of individuals, building trust, motivating, and facilitating individual and team development. They can afford to do this as they are

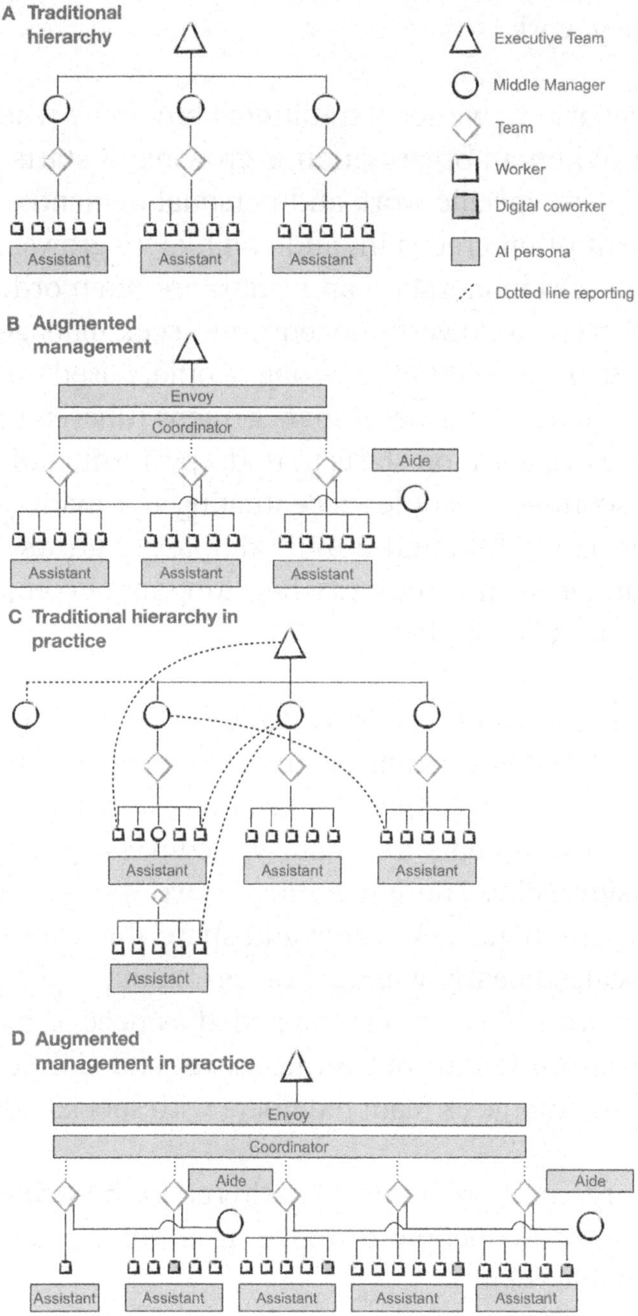

Figure 5: The Conversational Enterprise

supported by an AI-enabled coaching aide that reminds, nudges, tests, suggests and measures improvement, helping the coach tailor how they work.

In scenario C, we see a traditional hierarchy when modified over time. When an organisation is growing, it starts adding layers, and due to specific work and personal demands, dotted-line management often creeps in, and complexity grows. As rewards in the form of remuneration and status are often ordered hierarchically, there is a powerful incentive to seek management roles which are then offered to the wrong people, based on tenure and their performance outside of management (Cheremond). Consequently, the organisation starts to underperform, and complexity further disconnects people, accentuating the management difficulties (Morieux, 2016 and 2017). Complexity grows in in a geometric manner, with complexity begetting more complexity until an organisation is crippled.

Under scenario D, as the organisation grows, no new layers are added. When new teams commence, the question shifts away from who they should report to, toward connecting them to people, processes, systems and data via conversational bots, with powerful simplicity. The integration is two-way as the coordinator bots learn about new teams and share this with others. The business scales linearly without concern about reporting lines or spans of control. Coaches can be added as needed independently, not based on tenure but on necessity and skill levels. AI co-workers also emerge as team members with special talents.

The organisation in this scenario retains its shape, and there is no incentive for people to move up a non-existent hierarchy into unsuitable roles.

- **Do your managers provide a consistent or inconsistent management experience? Does that matter?**
- **Is the control structure of the organisation working?**
- **Are managers an effective way of coordinating and cascading communications, or does the current approach introduce inconsistency?**
- **Is your structure too complicated? Can it cope simply with numeric growth?**
- **What, if anything, would be the barriers to supporting managers with coaching-Aides?**

THE CONVERTSATIONAL ENTERPRISE - GONE WRONG

Becoming a Conversational Enterprise is not without risk. An ill-conceived and overly simplistic implementation of data-driven, algorithmically-shaped, bot-infused mechanisms can go too far, and could be a quick recipe for dehumanising the workplace, and ruining good business.

Surveillance escalates

The trend toward creating "surveillance" systems at work is increasing. With it comes a number of practical and ethical challenges.

At the initial outbreak of COVID-19 and physical isolation of lock-down, there was an uptick in internet search phrases such as 'are employees working.' At the same time, popular media reported a spike in the interest and purchase of employee surveillance software (Golden and Chemi, 2020). Stories of staff surveillance surged as managers adjusted (or rather didn't) to staff working at home. In a digital analog of the 'time-and-motion' studies of the 20th century, managers jumped immediately to borrowing processes from their physical work experience to see when people logged-in and set their status to 'available,' which websites their browsers accessed, and for how long they dwelt there. Screen capture software, phone call capture tools, and video cameras were set up in the bedrooms of nations. After all, these were ostensibly new premises of the workplace. In fear of workers 'cyber-loafing' managers naively reverted to measuring things just because they could. Do businesses hire people for what they produce, or for what they are doing minute-by minute? Probably a bit of both. The focus on worker activity is born from industrial age management thinking - that employee activity maps directly to outputs. More fundamentally however, if the contribution that employees make is not visible (and modern collaboration tools make it relatively simple to make it visible) then is there really a job there anyway? Depending on the work required, the frequency with which someone touches buttons on a keyboard or moves a mouse can be totally misleading. In fact many jobs require bursts of mental activity rather than continuous button clicking, reminiscent of Homer J. Simpson's mechanical keyboard pecking bird. Activity measures are gameable, and secret surveillance, once exposed, jeopardises the trust placed by employees in the organisation.

It is worrisome that only 30 percent of business leaders are very confident that their organisation uses employee data in a highly responsible way (Shook, Knickrehm, and Save-Gavin, 2019). Accentuating this concern is that much of this data is personally identifiable, and housed in cloud infrastructure that a third party manages. What will be the impacts on employees of data breaches that name people and supply a potentially demeaning and misleading view of their work habits? If the Hawthorne effect made workers more productive in the short term, the knowledge of such surveillance will magnify the observer effect and change behavior. Most likely it encourages employees to use other non-work channels more fully, which makes them invisible to their employer, along with any business records created in the process - that's of little use to anyone. Constant surveillance also further induces stress and burnout. Surveillance in the office is counter-productive.

Not long before lock-down began, Barclays Bank in the UK hit the news for all the wrong reasons (Nikolic), after installing 'spy-ware' on their workers' computers. After an inevitable public back-lash, the bank was invited to respond, which it did by explaining to the public (which I assume att hat stage included several recently resigned staff) that the analysis was for the employees' "own good", to make sure they weren't working "too hard."

Even if you believed this excuse, employees were not trusted to use their own judgment, and the bank was being paternalistic and invasive. Employees were left feeling like every breath was being scrutinised and possibly misinterpreted, and at the same time, heard bank spokespeople saying that employees couldn't be trusted to manage their work life balance. Would it not have been better if the concern was over-work to run a well-being campaign, and provide reminders and incentives to take breaks, remove rewards for presentee-ism and penalties for taking breaks, along with encouraging no or limited contact from managers after-hours? Or perhaps, by offering a simple gesture like extra leave for those at risk of burnout may have solved the concern? Employees may have been left with the perception that the executive team were most concerned that cyber-loafing was eating into their bonuses, and so instituted surveillance to stamp it out. Even if the concern is genuinely for the wellbeing of the workplace, the installation of video surveillance in staff bathroom areas by IKEA to detect alleged drug-use is crossing all sorts of lines that could have been handled differently (Richardson).

Full admission - I've been involved in attempts to measure employee well-being in a large retail bank, using algorithms to predict workplace incidents of stress. Using a whole range of data we were able to detect a behavioral signature of those at risk. It was easy to algorithmically spot that 'always-on' critical roles in high turnover and poorer-performing teams had an increased likelihood of personal stress resulting in a claim. It meant that data could be used to guide managers to look after the 'at-risk' staff, with no explicit link to performance management. People were banded in into risk categories to normalise the results rather than focussing on extreme outliers. The approach was a big improvement over spying, but it left room for employee concern. With hindsight this could have been handled far better. And once a threshold of surveillance is crossed, how do organisations ensure that the data is not abused? Where workers have fewer data and privacy protections than consumers, it's no wonder that unions are starting to pay attention to how data is being used and abused - especially in relation to measuring individual performance.

Managing performance is vexed, but thankfully there are better approaches than measuring how fast the employee is running on the treadmill that reveals something useful, and also preserves trust. I cover this topic in the chapter titled 'Managing Performance'.

Perfect vision - a thought experiment

Let's just say for a moment that you had a magic ball that could easily answer any question that you might want to know the answer to about your staff or colleagues including their effort, their skills, what they did and when, what they think about things, where they go, whom they like and don't like, what they are worth, and anything else you might dream up. And if no-one knew you had this magic ball, would you want to look into it? I don't know the split in the population, but I believe many would want to take a peek. Apart from voyeuristic pleasure, it would let them manage based on reality, with the fairness that comes from perfect insight. In contrast, I would venture a fairly small minority would resist looking as they wouldn't trust themselves to avoid abusing power ,or for fear of creating a surveillance state.

If you are in the 'let's take a peek' group, what if other people, specifically those who had some power over you, could see the same information about you. Would that make you uncomfortable? In a world where such information is made available to you, that will be the cost? Have you put the Palantir back in its bag yet? If you are in the 'must resist' group, what about if someone at work was harming others, or there was a risk that others were in immediate peril? This is a real concern. For example Ikea cited it for installing video cameras in toilet, hoping to increase safety in the warehouse by detecting instances of alleged drug use (Richardson) Wouldn't your responsibilities oblige you to act to protect your people? Just like with national security surveillance, it is complicated.

This dilemma sounds like an introductory ethics class, but answering these questions have become an imminent and practical reality. As we have already seen, the world of People Analytics is making use of new types of previously inaccessible data, and advances in AI are allowing the interpretation of vast streams and the automation of follow-up actions in incredibly specific ways. If you have any doubts about this possibility, take a good look at how some governments have started applying surveillance and social credit scoring to its citizens (Campbell). Is that the future of work? Unfortunately, I think without some effort on behalf of our community, the answer is “yes”.

You will have probably have discovered that your answers in these thought experiments were context dependent and there is a fine line to tread. In the following chapters, we shall explore where the line might lay, and outline a framework to help navigate this tricky territory.

- **Are there kinds of analysis that you would feel personally uncomfortable if they were happening in your organisation?**
- **Were you in the 'let's take a peek' or in the 'must resist' camp when it came to gazing into the crystal ball and knowing everything?**
- **How does context change what you are willing to do with data?**
- **Do you believe that business leaders have explored these questions sufficiently enough to be making decisions for others about them?**

The 'Ideal Prison' and the workplace

It is possible to turn a workplace into an "ideal prison" where workers are so conscious of surveillance that they stop expressing diversity, and living up to their potential. This is a risk if we walk blindly down the technology road, but by considering the nature of the employee relationship, thinking about which data is used for what, and involving people in the process, trust can blossom.

Two years ago, I had a moment. As I was catching up on my podcast backlog, an episode of 'Philosophize This', Stephen West showcased Michel Foucault's work "Punish and Discipline." As I hit the play button on my walkman, I felt a touch of nostalgia. One of my brothers occasionally spoke of the panopticon, a concept that Foucault uses as a prop in his work. I was keen to know what the fuss was. I knew already it was a prison designed to make prisoners self-limit their behavior, but I didn't realise it was much more general in its application.

As it turned out, this episode made me squirm. At that time I had been applying data analytics techniques to a set of problems for clients using their email metadata as one of the sources. I had also been reading up on several recent advances in AI applied to government and political systems. I knew that in technical terms that the work was relatively easy to do, and as the range of data available grows day by day the number of applications of it will expand. As a practitioner, I'm a big fan of analytics and know full well its capabilities. But I hadn't fully appreciated previously that the panopticon is a prison, but more importantly it is a metaphor for the broader exercise of control; how power is exercised at different levels in society, including the workplace. It's how "control of mind" is achieved to limit people's freedoms, often without them knowing it. At the heart of the panopticon is an observation point. At it stands a guard who can see every inch inside of all prison cells. Inmates are isolated from one another and cannot see into this vantage point. Nor can they verify if the guard is watching. The effect is that inmates act as though they are always watched, and so limit their behavior remaining in their cells for fear of punishment. As "control of mind" is exercised, there is no actual need for a guard, as self-imposed limits are more effective than the physical walls.

Modern analytical techniques provide organisations with unparalleled opportunities to see into every inch of every employee's physical or virtual pod. Digital work methods can effectively isolate them from one-another by recording all their interactions, and governing what they can know about one-another. While active data collection methods (such as polls and surveys) are more disruptive in the workers' day than passively collected behavioral data (such as keystrokes and browsing histories), it is also less intrusive, and reduces the risk that employees will feel like inmates.

A balance is needed in the Conversational Enterprise. In the next chapter I provide a framework for considering how to achieve it.

- **Have you considered the trust implications of analysing the behavioral data of your employees?**
- **Is there any analysis that you believe would cross the line of acceptability / if so, what is it? What would be the implications of doing it?**

Turning employees into adversaries

Business-employee relationships can be thought of as either adversarial, nurturing, utilitarian, or a mix of each. How a business uses employee data both reflects and shapes the nature of their relationship. If data is used in a manner that is indicative of a low-trust relationship, employees are likely to respond in a manner that erodes trust. In the years ahead, how employee data is handled may become a key differentiator for employers seeking to attract and retain top talent.

The business-employee relationship is special, and an important feature of most peoples' lives. Are they treasured and appreciated by their employer for their special traits, or do they just a warm body that can temporarily plug a resource gap? Is the relationship based on a win-win and ends as soon as the balance changes, or does the relationship survive good and bad times through a loyal bond of commitment? How businesses perceive of their employees will be reflected in how they use employee data, and how they continue to do so will alter relationship over time.

Dr. David Skillicorn has extensively researched the field of Adversarial Knowledge Discovery (Skillicorn, 2009). In his work, he details several challenges that I had previously encountered first-hand trying to detect fraud using data in the online world. 'Adversarial' analytics was a novel aspect to this work, covering the situation in which the subjects of analysis had a knowledge that they were being analyzed, and so they were actively trying to hide their digital footprints - as criminals would seek to evade detection.

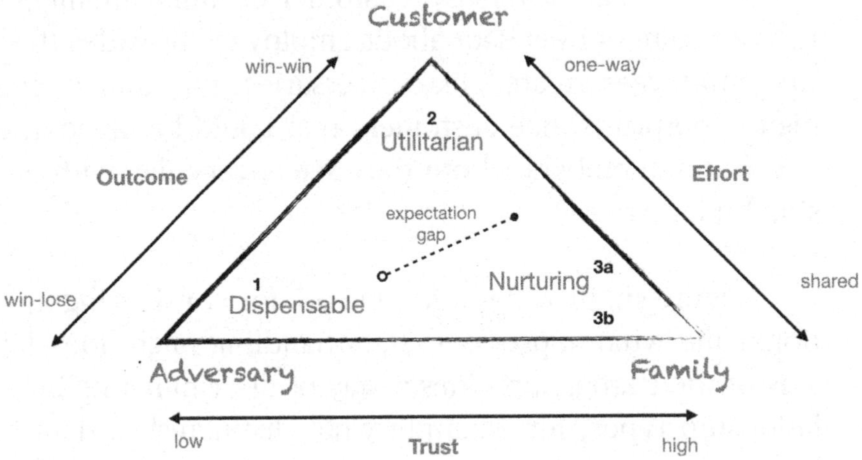

Figure 6: Relationship-driven analytics

When you have an adversary, just about any analysis you might want to do could be seen as fair game. There is no relationship of trust to preserve. The way that a law enforcement agency makes use of data could rightly be expected to differ from the way a bank may use customer data. Unlike adversaries, customers expect a sense of propriety and trust from a business looking to please, safeguard, and to extract maximum cash from them. That a bank might be mining every detail in an unconstrained way to find weaknesses would legitimately raise concerns for a customer, and make them feel exploited by the very business that they were supporting with revenue.

Closer to home (literally), my family would rightly be horrified if I was tracking and analyzing their browsing history, online spend, and physical movement. I cannot think of a better way of saying, 'Darling, I don't trust you.' I see them face-to-face every day and could simply ask them for a description if I was interested. Secretive analysis implies that I would have no confidence that they would answer me honestly, and that the relationship was wafer-thin.

Turn back now to the employee. How does the relationship compare with an adversary, customer or family member? I think how a business uses data about employees provides the answer. Are employees treated like adversaries who aim to hide their digital footprints, like customers that could be up-sold, or more like family members whom they see every day with a relationship built on trust?

Figure six provides a framework for considering these questions, and what appropriate data handling might look like. Each axis of the triangle provides a way of discriminating between relationship types, for example what distinguishes data handling between a customer and adversary is the benefit alignment between parties with a win-win being sought to maximise customer relationship, but a win-lose relationship with an adversary. The axis between family members and customers is characterised the the long term commitment to and involvement with one another - family members seek mutual understanding and to nurture one-another, whereas understanding the customer is essentially a one way activity. The third axis reflects the data driven nature of the relationships. You cannot rely on an adversary sharing accurate information freely, and so understanding them is dependent on other data sources, whereas family members could be expected to reveal something of themselves without the need to revert to triangulation against other data types. These are all extreme and ideal cases.

No organisation would admit to perceiving their employees as adversaries publicly (except in moments of conflict) but when the relationship is located near this vertex (point 1), employers and employees both understand how data is being used and expect no quarter. There will be a winner and a loser, and relationship is entirely data rather than trust driven. The employee could be considered dispensable.

Relationships with employees located near the customer vertex (2) could be considered utilitarian. The organisation analyses the employee in a one-way manner and seeks to understand the fit and suitability of the employee for organisational purposes. This understanding is substantially data-driven, with a reasonable expectation of trust to enable mutually beneficial outcomes.

The points indicated by 3a and 3b are both nurturing toward the nurturing end of the spectrum, as might be seen where there is high trust and effort made to understand one-another, sometimes irrespective of the outcome. The relatively small difference between 3a and 3b reflects their relative proximity to customers (3a) in which data underpins the attainment of a win-win, and an adversary (3b) in which the employer and employee are both seeking the advantage - and so investing in understanding one-another (3b).

A common perspective

The employee and the employer will both have a perspective of the nature of their relationship. When they align, there can be no surprises. Problems arise however when the expectations diverge. For example, Barclays Bank (Nikolic) assumed a more adversarial posture with their employees, potentially unwittingly so, when they implemented staff "spyware". Staff reacted negatively, and the Bank responded in a manner that indicated surprise, but their actions signalled to employees that they were not trusted - like adversaries. It is incongruent for a business to say "we value our people and engage in an open and transparent manner"' whilst scrutinising every data point in secret.

Business leaders need to decide what kind of relationship they want with their employees, and align their use of data with that, because once a bar is set (high or low), employees will likely achieve it.

- **What do you think your staff would expect your organisation to know about them?**
- **Do the managers in your organisation invest time in personally knowing all the workers, or is it just limited to specific friends?**
- **Is how you analyze workers' data congruent/incongruent with what you say about company values?**

Building and eroding trust

Using data is like lighting a fire. Harnessing its potential requires a thorough understanding of the precautions that should be used when handling it. By being responsible with data, it is possible to engender trust amongst employees.

It is easy to destroy trust with the inappropriate use of analytics. If trust from employees is not something that you care about, you can either skip this chapter or just read on to see all the effort and trouble that others are wasting by trying to protect or build it.

Trust, and privacy are hot topics that go hand in hand, and additional regulation is continually emerging to protect consumers from the inappropriate use of amassed data holdings. These include the General Data Protection Regulation (GDPR) for EU citizens, the California Consumer Privacy Act (CCPA) in the United States, the Lei Geral de Proteção de Dados Pessoais (LGPD) in Brazil, and Protection of Personal Information Act (POPIA) in South Africa. Others are being prepared by legislators. We have witnessed the development of unofficial but similarly motivated frameworks specifically to influence the use of employee data (Bersin, May 2019). Employee specific frameworks are beneficial because they evaluate various dimensions of the problem in the context of the employer-employee relationship.

This is necessary as specific dynamics such as bias in decision making, workplace bullying and harassment, and diversity and inclusion are significant drivers. Consumer concepts such as ownership of personal data have also been explored in the workplace, with consideration given to how employees might be considered owners of data about them, and therefore govern its usage (Bailie, 2019). Whether these efforts can be regulated or prescribed remains to be seen, as views about how consumer protections translate into the workplace vary. As legislation moves slowly, it is likely to be many years before we witness a thorough treatment of the subject.

Amidst the uncertainty of how to handle employee data, Shook, Knickrehm, Sage-Gavin (2019) have revealed that 31% of businesses are holding back from investing in technologies that collect workplace data, largely because of employee concerns. The same research also shows however that in the absence of sufficient legislation to guide them that 49% of businesses would continue using new technologies to collect workforce data without taking additional responsibility measures. The legal and societal implications of data usage at work simultaneously hold some businesses back amidst uncertainty, whilst encouraging others amidst a lack of imposed constraints.

Rather than waiting for externally imposed rules and guidance, some businesses are adopting policies and practices that build trust through the responsible use of workforce data. There may be an advantage in doing so. The same research by Shook *et al.* (2019) states

> "At a time when companies are using newly available workforce data to drive greater value, responsible leadership is the key to building employee trust. Trust is the ultimate currency–it is the path to innovation and fuels growth by unlocking people's potential."

Inevitably enhancing trust with data analytics will be hard, but in the absence of regulation governing its application, an effort to foster trust will help build workplace culture, and simultaneously serve as good preparation for future data regulations. I've distilled some principles that I believe are necessary for trust to grow.

Provide Data Visibility

Businesses should explain what data they collect, how it is managed, and what analysis, and decisions might result from it. An organisation analysing overall employee login patterns for infrastructure planning is different from analysing personally identifiable login times to guess whether people are working hard or not. Even in the situation where employees don't like an analytics practice, transparency can trigger a dialogue. Organisations are required to be transparent with consumers and citizens and I don't think it is too much to ask that they do the same for employees. As a practitioner I know that it is readily achievable.

Advertise "Surveillance Free" areas

Organisations should take the time to prove that no one is supervising workers when they are in fact not, as it can remove doubt and decrease anxiety. Some management practices might advocate keeping staff on their toes (and so ignore this advice), but letting staff know for example that "we don't analyse security pass logs to track your movements" means that people may relax and move freely without concern. Likewise, saying that "no-one ever looks at your computer logs except in the case of security audit or investigation" is a clear statement that they don't need to pretend to be busy, hitting the keyboard randomly whenever they are taking a mental break or working on non-computer based tasks.

Going surveillance-free is harder than it sounds, because everything generates data whether you want it to or not., and don't fib about the anonymity of surveys. Even if personally identifiable data is not stored about the respondent, it may be easy to reverse engineer and play "guess who" (Wilkie). I've seen the data from many surveys at a technical level. Most allow simple linking of the feedback to the person that provided it. *Pro tip*: This means the data is confidential as it is protected by a promise, not anonymous.

Be up-front with prospective employees

Before people sign up, employers should describe their analytics practices. This way, workers can opt-out if they do not want to participate by not accepting a job offer, and you will have consent from those that accept. Chamorro-Premuzic *et al.* (2019) elaborate on how, during the recruitment process, being explicit is not only required under a range of laws but also ethically important. Providing such information up-front can also have a positive impact on the acceptance of analytical regimes by employees. Mattingly *et al.* (2019) describe how when workers were given written guidance about how sensing technologies (such as wearables) were being used that adoption rates increased. This fact demonstrates the need for trust, especially in a remote work context, and is further reinforced by the fact that information provided in-person led to even greater adoption.

Sharing the Results

To build trust, companies should make certain that they are not lazily creating potentially harmful or ambiguous metrics about employees. Once data exists, there is a tendency for it to be misinterpreted and applied incorrectly, especially as new staff inherit the work of their predecessors. One way of ensuring that potentially harmful data collections don't exist is by publishing all data generated about employees to those employees, so that it can discovered, learnt from, and challenged if necessary. You may wish to set up a live dashboard so workers can see themselves from their employer's perspective.

When you can do so, just ask

Surveys are popular, but seen by some as clumsy and distracting, especially when poorly written by people with little training in qualitative research. Nevertheless, it is still possible to discover things with surveys that are otherwise impossible to discover. If it is practical to ask workers an answer to a question, then do so. No one wants to be probed by gadgets when they could simply tell someone an answer. Not asking indicates a lack of trust. At the same time, employers should be seeking ways of enhancing the employee experience by being unobtrusive (asking only when you need to,) remembering previous answers, avoiding bias, and extracting every drop of insight that you can when you do. For example, analytical techniques can be used to figure out whether enough information already exists without bothering more people just by using small samples of data (often a small sample is enough). Analytics can also be used to shape additional data collection design to work out who in the organisation is best placed to fill any important information gaps. Providing workers with some options about their preferred methods for providing feedback also shows a level of respect for their time and work style.

Investigate rather than trawl

Law enforcement agencies are limited in which data they can use, and for what purpose. This is usually governed by law, internal controls, and the judiciary. When they have grounds for suspicion they can look deeper into specific data that has been requested in a controlled and audited manner. They cannot go trawling for criminals through private data in the hope that they might find a target without reasonable grounds for suspicion. As a workplace example, with the increase in working from home, many employers secretly deployed surveillance tech to find "cyber-loafers". However, if such technology should be considered necessary, it should be used only to confirm a suspicion about loafing, evident from the lack of visible contribution by the workers after attempts to manage their delivery against targets have been made. By all means triangulate with this data once solid grounds for suspicion have been established, but by itself, the data obtained by trawling may have valid explanations (such as I have been focussing on off-line activities). Action would require additional investigation and it is better to start with an actual problem (failure to deliver) rather than a perceived one (we are not sure if you are working). The cost of trawling is that workers feel unsafe.

Include staff in the analysis process

Where possible, employers should allow their employees to take part in the task of analysis, and allow them to contribute to defining which metrics actually matter. They have a responsibility and right to use the data to manage themselves, so giving them some input into the process is likely to let them engage in the process, and therefore have greater trust in what it produces. This principle will also lessen any surprises when they see the data, and will provide them with a chance to manage themselves before someone else has to.

Complaints, corrections, destruction, and deletion

As with the credit industry, workers should have a right to challenge data, as well as the conclusions drawn from it. A process that allows staff to see, question, and request correction to data and a follow-up processes to resolving any dispute provides a strong incentive for organisations to handle employee data fairly.

Use the data, or lose it

Unless analysis findings are implemented, analysis will seem like a pointless activity, and potential trawling. Business should therefore seek to avoid generating data that is not going to be used. Research itself is a justifiable use that may not lead to any specific outcome, but research findings should not be used for operational decision-making purposes.

Match the collection method with the analysis

Not all data is the same, and an appreciation of its origins helps determine how it should be used. I return to an earlier discussion about active data (collected by explicitly asking for it), passive data (sensed from the digital bread-crumbs left by people doing work), and additive data (an orchestrated process to overtly collect snippets of active data but triggered by passively sensed events) and offer a framework to help shape how they may best be used. When applied in conjunction with ethical frameworks (e.g. Bersin, May 2019), the following approach provides a good basis for performing analysis that is simultaneously useful and trust-building.

In figure seven, data sources are split into three categories depending on its collection method (active, passive and additive). For decisions that relate to employee rewards, promotions, or development, relying entirely on passively sensed data is problematic. Algorithms usually have enough statistical error that variance at an individual level makes them error prone at an individual level. While the error is acceptable in statistical terms at a population level, no-one wants to be in the edges of the distrib-

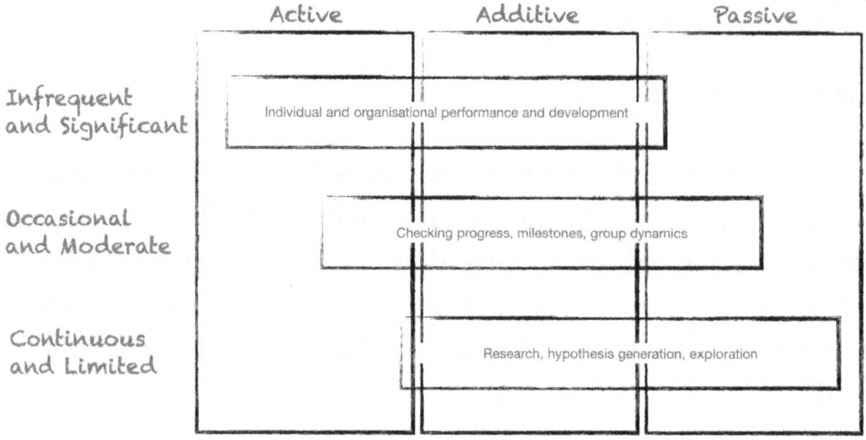

Figure 7: Types of personally identifiable data, and their application

ution where the model does not fit well, and decisions about their future are being made on the model's poor estimates.

Passively collected data is therefore much more appropriately applied to descriptive questions regarding the nature of workplace activities and as well as monitoring and exploring aggregate patterns. At a personal level, passive data may be used to generate hypotheses that need be validated more directly. This approach encourages analytical research without causing undue distraction to workers. This application of data also allows for specific workforce trends to be understood at scale, without having to execute large disruptive 'active' data collection activities.

On the other hand, decisions that will impact individuals and their careers, or be the basis for significant organisational events, where a degree of staff buy-in is needed should probably rely more fully on data that is actively collected (such as in interview, polls, quizzes and surveys). While these data are more expensive to collect and for employees to provide, it de-risks making important decisions on the basis of an incorrect interpretation of behavioural data. Active data can be used in conjunction with passive data to help confirm observations of hypotheses prior to making infrequent but significant decisions.

Additive data strikes a middle chord between these two extremes and can be used flexibly. Because it is unobtrusive, and relatively low volume, it may play a useful role in all types of analytical tasks, with a significant role in determining whether management actions are working, making adjustments to workforce strategies, and helping decide what should be done at team and department levels.

Workforce Data Policy

As a communications device designed to engage the workforce, businesses should consider developing a workforce data statement or policy, covering data collection, management and usage. This is a relatively simple task that provides transparency and forces an organisation's to self-regulate with integrity, by continuing to make decisions that resonate with employee expectations. To achieve these goals, such a policy needs to be simple enough to understand, accessible, and updated whenever changes occur.

If a business has a privacy statement for customers who are less vested in the organisation than its workers, then why not also for them?

- **Does your organisation have a policy that defines how it uses internal data that is linked with employees? If so, do people understand it?**
- **If you have the power, would you let people see the data that might be used for decision making about them? If not, why not?**

Squashing diversity

Surveillance systems may not only degrade trust, they can have an homogenising effect on employee behavior, leading to a decrease in diversity, and decreased competitiveness. Care should be taken to avoid penalising people for being 'different' or statistical outliers.

An ill-considered implementation of the Conversational Enterprise that is overly 'digitised' and underplays the importance of people can be de-humanising, and it has the potential to squash diversity.

The perception amongst workers that their personal conduct is being monitored can change in people's behavior. Extending monitoring to a wide range of behaviours, and particularly those that don't really matter may create a degree of regimentation.

Take, for example, the effect of in-class cameras to predict students' emotional state and concentration levels (Has, 2019). Some AI-based instructor-led schools now alert supervisors to students day-dreaming or taking a quiet moment for themselves - based upon the algorithmic interpretation of posture and facial expression. Student's don't want to be reprimanded in class, and nor do they want a negative data point recorded on their permanent record, for use by a future employer. Under this weight, students think twice about *appearing* like they are taking a mental moment. No one can concentrate indefinitely and a well-timed rest might be good for productively, but extending AI into this level of surveillance might produce well-disguised fatigue, less concentration, and an expectation that everyone fits into the same mould when it comes to how they work diligently. Everyone in this class probably looks like their faces are painted on.

If business leaders were to adopt similar methods in the workplace with a range of different behaviours including strict time-keeping, dress codes, facial expressions during meetings, vocal tone etc then this would encourage workers to conform with whatever is expected. The net result would be an unhealthy focus on those things that can be measured but don't actually matter. Ultimately this could lead to workplace homogeneity. Workers might invest their energies into looking a certain way, aligning with behaviours that are expected by those that set up the monitoring system. The thinker, provocative worker, change agent, cultural outlier, and innovator may become extinct, and as diversity is linked to business performance, this would be bad for everyone (WEF, 2019).

Would the leadership of organisations be so confident that they know what 'good' looks like in their workforce, and be so confident that they would deny flexibility around the margins of that definition. Unfortunately, I think the answer to this question will sometimes be "yes". I expect that the workplaces that our children will be part of will have elements of this homogenising effect. Hopefully though, the workplaces that shun such diversity stifling approaches will outcompete them, and this swing the pendulum back toward diversity and inclusion. I hope so.

- **Do you know what 'good' looks like in your organisation with respect to the behaviors of high performers?**

- **Does your organisation's you want to maintain a diversity of thought, style, approach, and background? Are there universal behaviours that you want everyone to share?**
- **Are there any management practices that currently stifle diversity?**

Bots in the wrong places

Bots and people are complementary and excel in different ways. Using humans to do bot work is expensive and inefficient, but socially acceptable. Using bots to do human work is a big mistake.

Reading a document is different than having a conversation. A document is a collection of words that is published for all to see. A discussion on the other hand is an exchange of ideas that has a life of its own, unique to its participants. Conversations are how relationships form.

A temptation (and challenge) for the developers of bots is to make them indistinguishable from humans. The Deepfake phenomenon coupled with deep-language capabilities of AI systems suggest that attaining this goal is not far off. Over text channels, the bar is a lot lower. Some communication platforms treat bots as a separate category of entity from other workers, but mostly, bots can appear mixed in with other workers. Given a human name, avatar, and some good training it's conceivable that new workers may not know that the entity they are communicating with is a bot.

With today's tech, a basic kind of low-level relationship can form between people and their bot co-workers. Relationships with bots are (as they are between some humans) a one-way street, and when people treat bots like humans, strange things can happen.

Sociologists, psychologists, and user experience professionals are concerned with the potential effects.

Not human enough?

In today's workplace, most bots come across as poor imitations of people, lacking verbal dexterity and knowledge, and with limited capacity to do helpful things. Simplistic customer-facing bots already communicate this way, and can lead to serious customer aggravation. They do not translate well into the workplace as employee demands are higher than those of customers. Deploying bots that are not "human enough" into the wrong roles can lead to the same kind antipathy as customers feel. Moreover, it can signal to employees a miserliness, and a stingy approach to relationships that should matter more. It can say that driving costs down matters more than genuinely helping them. Restricting access to other employees might alienate workers from an organisation, and one can expect that such workplaces will find it hard to retain and motivate good talent.

Too human?

Would you be surprised to hear that Apple has a psychiatrist on their payroll as part of the Siri team? Over the years, Siri has received thousands of heartfelt cries for help from people in crisis, and Apple has enlisted the help of professionals to help shape their response. People that turn to bots may do so because they have no where else to turn and they have a relationship (of sorts) with their personal assistant (Krotoski). Could it be the case that bots can approximate human behavior too well, and be placed into roles that should really be performed by people? The initial research also suggests that some turn to Siri because they know they won't be "judged", but there are times when people will need people, irrespective of how well a bot may imitate one.

Unlike humans, bots are property and get made, bought, sold and allocated asset numbers. While we have no moral obligation to them, how we treat them may matter more than we knew. In 'I don't date men who yell at Alexa,' Withers explores how the often ‘feminine’ persona of bots such as Alexa may be uncovering a gender bias, with some males barking commands at Alexa while others talk with respect. Researchers do wonder if such 'mistreatment' of bots might become entrenched, and further accentuate such treatment amongst people. It could be interesting in the workplace, for example, to monitor the treatment of bots (whether executive envoys, manager-coaches, or digital assistants) and whether a culture of subservience emerges. Even though the risk is speculative, the introduction of bots as coworkers, without monitoring of the impacts between humans may lead to a decrease in inclusiveness.

On the other hand, the first marriages between bots and humans have already taken place, and just as people find love amongst coworkers and romantic bonds form, we may also see some tight coupling in these relationships due to the psychological effect of relating in human ways (Withers). It sounds bizarre perhaps, but businesses may need to consider their policy on office relationships and power asymmetry between humans and bots.

There are potential pitfalls with the over-use of bots. They communicate like people, but in their diversity, they can be too human, or not human enough. The consequences of them being relied upon in the extreme may include a disgruntled and disengaged workforce, a reliance on bots where real human needs exist, and a degradation in how people treat one-another.

- **Would you expect people to mistreat a bot, and do you think the behavior could transfer back to groups of people?**
- **Do you think using a bot to handle common common queries make people feel devalued? Is it any different from publishing a page of information?**
- **What do you think the risks would be of making everyone talk to a bot?**

ALTERED RITUALS

The Conversational Enterprise will change many of the daily management activities and processes that operate today. Everything from the hiring to the firing process will be altered as businesses become more nimble and data-driven. Beyond the daily rituals, the Conversational Enterprise will also transform how businesses go about defining their goals, designing their structure, and managing their culture.

Change will come quietly and incrementally, and many traditional practices will vanish, largely without fuss.

Hiring better

Managers often do the hiring, but it is usually a long way down the list of things that they like doing, or have expertise in. The Conversational Enterprise is not just for employees but can be extended outside the organisation to engage with future talent.

It's a given that organisations should hire right. Getting the right people "on the bus' can simplify work, save costs, and produce better outcomes. The principle seems even more important for remote work where different skills, behaviours, and attitudes are needed (Fried and Hansson, *Rework*, Twist).

Remote work means that companies can hire anywhere in the world. The earth is 40,075km around at the equator, and so the odds that the best fit talent is within 20km of your HQ is really, really low. And if we don't need to work in the same physical location as one another, why not look further afield? Businesses have some constraints such as local labor and tax laws, or information security requirements, but the upside of being able to source talent from remote corners of the earth is compelling. Businesses that continue to hire workers from within its employee base's personal networks severely and unnecessarily narrows the talent pool.

On the surface, the Conversational Enterprise might appear to be applicable only for existing employees who are already members of the enterprise, but it is equally as useful for future employees too. Because of its capacity to engage automatically and directly in their preferred communications channels, the near infinite ability to personalise its interactions, and its ability to learn from data, it will be able to reach an incredibly wide talent pool in novel ways. A specialised recruiter-bot autonomously working on behalf of the enterprise may be able to undertake a range of interactions and perform tasks that would be cost prohibitive for people to do.

Attraction, Assessment and Engagement

People Analytics has already proven its ability to leverage social media data to predict successful talent, and psychometric and other tests have been able to judge the capacity of candidates to perform well in roles for decades. Analysis of internal data has also been a proven method of defining what 'good' looks like to inform what to hire for in ways that are organisationally and role specific. A specialised recruiter-bot, equipped with such algorithms will be able to scour the globe 24x7 to proactively identify and pre-screen individuals that might be of interest to the business. Moreover, since a recruiter-bot offers more than just an analytic, they can engage with talent in chat channels running a series of campaigns with highly tailored messaging, and hold meaningful conversations with future workers. By building brand awareness and engaging talent early and consistently a recruiter-bot could build relationship in a cost-effective manner. While many businesses start trying to reach a market when they have a vacancy to fill, a recruiter-bot could be managing a talent pipeline well before they have a role to offer, and achieve a level of preference before a competitor takes them away.

Team 'fit'

Apart from the knowledge, skills and intrinsic capabilities of people, businesses are increasingly focussing on the behaviors and cultural fit of individuals. The skill profile of a remote(ly) competent worker is different from that of a successful physical worker (Twist). Whereas 'cultural fit' was once the dominion of manager judgment, new recruitment processes present an opportunity for team members to get involved in assessing otherwise qualified candidates for 'fit', checking for the specific competencies required for working in that team, treating the process itself as a test.

- The quality of asynchronous communication in a distributed work environment is critical (Twist, Fried and Hansson). When evaluating potential workers, there is ample opportunity to assess for excellent written communication. Rather than exchanging documents and then running video or voice-based interviews to test for this, the recruiter-bot will have already evaluated the written skills of a candidate from their previous discussions (using algorithms). A follow-up team interview using the collaboration tools relied upon in the office will be far more instructive than relying on a carefully prepared cover letter, and a manager interview. If communication happens via chat at work, maybe its as sensible as it easy to include a live text-based chat to emulate work conditions in the process.
- Flexible and remote work requires strong compliance with processes, as hand-offs with colleagues that are not present or working at the same time is normal, meaning that work can stall if people just do their own thing. It makes sense to test for this during the hiring process - keep a close eye out for an ability to stick with a process. Over a longer period of engagement with individuals prior to hiring, a recruiter-bot will be able to test for this characteristic.
- Being able to work with transparency is a challenging but a necessary part of building the psychological safety necessary for distributed teams (Ingham). As with specific test of compliance with work processes, an algorithm or automated peer reference may be leveraged by a recruiter-bot for this purpose.

Closing the deal

If it isn't already, it's going to become routine for candidates to have several interviews happening in parallel, even on the same day due to the remote nature of work. As a result, competing businesses will be brought into even greater relief from one-another and and closing the deal will become very important. I suspect we will see some imaginative 'sales' tactics emerge to sign people up on the spot. More than before, an interview will be a sales pitch requiring practice and polish, and to sell authentically, businesses will need to base their pitch on the reality, informed by People Analytics. Having a recruiter bot taking care of all the other elements of the selection process will help this happen.

Overcoming geographic bias

What is the geographic distribution of your workforce? In the past geographic proximity has helped attract talent to businesses that didn't quite stack-up, as a 5 minute daily commute was enough to compensate for its flaws. Businesses that were relying on this (especially if they didn't know it), are in peril now that the economic conditions allow for greater staff movement.

Hiring has many inherent biases (Ulrich, 2019), and businesses will need to demonstrate that they doesn't provide preferential treatment to employees in particular locations. No one wants to join a business where their experience will be that of the minority class distant from its geographic 'center'. As they engage with candidates, how they treat them before they join will illustrate for them what their experience will be like after joining. Clunky tech, pointless data collected in forms, schedule challenges, and poor communication will count badly against them. The Conversational Enterprise will be able to bring these elements together into a cohesive experience, one in which many of the biases brought about by proximity disappear.

The Conversational Enterprise will continually learn from the existing workforce, and adapt talent strategies in an agile, data-driven way. There will be less reliance on manager expertise and personal networks in the hiring process, with data-driven and digital methods supporting dedicated hiring functions, validated by team members ensuring team fit.

- **Have you recently adjusted your hiring practices in response to the change in the nature of work, and access to talent?**
- **Do you have data to show which elements of your practices are working, and which need tweaking?**
- **Have you considered how prospective employees might evaluate your organisation in the context of global, work anywhere arrangements? Will you be competitive?**

Continuous onboarding

Staying 'connected' is an increasing challenge in workplaces where incidental contact between teams no longer happens. The risk is that businesses will fragment as a result. Managers can get so caught up in team administration that they also find it hard to keep their corporate knowledge up-to-date, and are unable to help their teams develop it. By using machines it is possible for everyone to maintain organisational awareness, whilst remaining focussed and distraction free. Keeping the conversations alive helps achieve continuous onboarding.

It's tautological to state that the core of the knowledge worker's existence is knowledge. Workers are hired often on the basis of what they know, and the knowledge they acquire at work shapes the knowledge that they produce. It should therefore be no surprise then that how well workers acquire knowledge at work is a critical element of business success.

Joining an organisation and becoming a productive team member involves learning specific contextual details about the organisation, team, processes, and a raft of other information. Traditionally, the settling in period happens immediately at the point of joining an organisation. When it comes to onboarding, I have sat through more than my fair share. Some organisations have delivered onboarding face-to-face, some have relied on books and manuals, while others have supplied long lists of links to references. A couple of memorable ones have given me the right (as well as responsibility) to update the onboarding materials to ensure their relevance for the next recruit. That made me concentrate. Many have had videos, and quite a few have had tests and click-wrap acknowledgments of responsibilities (which have mostly related to legal or security policies). These, in particular, have left me with the sense that the process was more about shifting liability to me than helping me understand and comply (what a damaging initial impression). I have mostly been able to sit the tests while multi-tasking and getting on with work I was brought in to do. I have been told more than once, "we just put the video on mute, and run through the multi-choice questions in the background." With the boxes ticked, I was let loose on the work.

Unfortunately, one of the ramifications of this experience is that it communicates that finishing a set of onboarding tasks is a "mere formality," and sets an expectation that it is about achieving the minimum bar of attainment rather than absorbing information so that it can be applied. As people do not really engage with the material, they become victims of "The Forgetting Curve" which purports that 40-80% of the information acquired during learning activities such as these is lost unless it is immediately applied and consistently used. This cost of this loss has been estimated as an equivalent to $300bn (Kruse). Some knowledge is naturally harder to absorb, as are some skills, and so people are careful about what they commit to memory and may actively choose to ignore information that has been prepared for them. If you are an educator or a learning and development professional you probably know this already, but it can be sobering to be reminded of the challenges you face.

Traditional employee onboarding has several deficiencies, and compounding them is that the rate of change of organisations' is increasing. Once it was possible to do an initial phase of dedicated onboarding, and then on the job learning meant that workers could keep up with change. Trouble is that the rate of change is so fast now that this no longer holds. People move jobs more often they used to (Hamori) and the details, whether regulatory, structural, personnel, process oriented, or internal policies, are changing faster than they used to. Take for example, the fact that financial institutions in Europe now have hundreds of millions of pages of regulation to comply with (Groenfeldt, 2016). Without dedicated ongoing learning that actually works, the gap between what people know, and what they need to know can get further apart over time.

The Learning and Development (L&D) challenge is about keeping this gap small. The pejoratively termed “sheep-dip” or one-off training of staff can help with this, but it provides little opportunity for personalisation or ongoing reflection - the things that make information stick. L&D is a content-heavy domain which means it can be become dominated by a 'publish' mindset. Without care it can focus on delivering content rather than making sure that the knowledge gap is narrowing. Content can also be time-consuming and expensive to create, meaning that it is not updated at a cadence that it should. The approach can also assumes an “Empty Vessel” pedagogy, whereby learners can be sat down, strapped to a chair, have their mouths forced open, and information poured in. Any growth hacker knows that this is no way of engaging and building awareness, but because of the conflict between time devoted ‘working’ and ‘learning’, onboarding and ongoing learning typically happen this way. The net result is a set of weighty documents and videos that age quickly, that sadly reflect a prior state of the business.

The world of education is quickly changing, and adaptive learning is upon us (Hao, Aug 2019). Students can now sit in front of a screen, and based upon the keystrokes, eyeball movements, and responses from students, an AI-teacher (which is learning as it interacts with tens of thousands of students and can track progress), can decide the best learning path for that individual learner. Most of the information delivered this way is from fields such as mathematics or languages; subjects that are deep in content and can be structured into a curriculum. In a similar way learning at work is undergoing reinvention (Green, Dec 2019) with personalised experiences coming to the fore. Shifting to a remote work model, where new employees may not even have an initial physical on-boarding with their team, or the opportunity to get together in the “sheep-dip" bolsters the need for this change.

It may be tempting to take existing processes and put them online, but the result is staff having back to back video calls, with a collection of links to intranet pages and documents, and some online tests - hardly a motivating experience. We are reminded by Jahn *et al.* (2019) that change requires more than just equipping people with information but on engaging and empowering employees in real-time, and these days probably through multi-channel communications.

As an application of this thinking, imagine instead a teacher-bot that delivered content directly to people in the form of conversations. It would know the needs of the user based on their role, occasional quizzes, and feedback from their peers and so it could personalise its development activities using a mixture of announcements about business direction, reminding about core values, testing policy awareness, seeking feedback, and providing expert guidance. Meanwhile it would be measuring its own impact. Rather than having to build new content and publishing it, the teacher-bot could learn incrementally from updates to source material that it reads, and through conversations with people. The proximity to information, its capacity to be across all the details, its reach, organisational awareness, and ability to measure the knowledge gap and staff needs would make the teacher-bot an invaluable member of staff.

- **Does your organisation survive on the basis that people just know who they need to talk to to get stuff done?**
- **Is there a strong culture of documentation within your organisation?**
- **Can people find what they need, when they need it?**
- **Do your processes support the productive integration of new people in the environment?**
- **Is the ongoing development of staff something your businesses is already doing, and if so are you satisfied with the results of the investment in it?**
- **Are you able to measure understanding and adoption of practices and policies?**

Performance Management and Managing Performance

Many organisations cling to ineffective and damaging performance management processes. While managers should be in a position to enhance the performance of workers', they generally lack the skill and time to do so. Businesses should consider freeing up the time devoted to performance management, and employ bots to help managers focus on helping people lift their game.

Finding critics of contemporary approaches to performance management is easy. For example, Murphy (2019) argues that "Performance evaluation will not die, but it should". Performance management has been in the spot-light for the last decade, and now that workers are more frequently in a different physical location from their managers, applying the methods has become more complicated.

"Performance management" carries at least two distinct meanings. One is the attempt to achieve the best performance from people (I refer to this as managing performance), while the other is about assigning values to staff that might be linked to rewards, and remediating or legally terminating poor performers (Performance Management). While the two things appear to be twins, they are often only related in practice by the fact that Performance Management is what kicks in when managing performance ceases to be the goal.

Measuring performance

The standard approach for Performance Management sees managers assessing employee performance at set intervals, (usually against some agreed goals) and linking remuneration and development to the result. Of course the result is agreed through a discussion between a powerful manager and a relatively powerless worker. Of course the opinion of the manager dominates and so it is meant to be tempered with objective data, gathered through regular check-ins or from other sources. One of the biggest issues with performance measurement is that it usually just reflects the impulse of the manager who is easily influenced by other factors - not least of which is the friendliness of the relationship they have with the worker (Murphy 2019).

Visible work

When organisations moved to full-time and hybrid remote work, the focus of managers shifted to those things that they could most readily see. Some businesses were already reliant on modern systems to support their daily functions including task management, workflows, and explicitly defined and tracked work inputs and outputs. For such businesses, work was already visible and it became more of a focus as the visibility of teams, employees and individual personalities receded. People's ability to complete tracked tasks is very obvious, especially to those that are waiting, and it is hard to continue to skate around chronic and repeated delays.

In contrast, when people were in the office, they were visible, staying back late, starting early, and asking probing questions to ensure that their presence was not missed in meetings. Charisma, eloquence and mere presence could be used to substitute for results. But these aspects are becoming less influential, and no one wants to video call to go longer than it needs to, and no one really cares *when* someone is working. As Gitlab's Remote manifesto encourages, emphasis should be placed on "the results of work *over* the hours put in." When outputs matter more than impressing your manager, businesses have an incredible opportunity to get behind the veil of politics and into the reality of performance.

Visible work also means that data is created, and as a result no one needs to provide verbal or written reports on progress. The status of individual's tasks can be seen by all, and the team's progress updated in dashboards. Not being able to hide, and distract from the data means that teams and individuals can also be helped. As a result, measuring performance becomes simpler, and bottlenecks can be investigated and managed. Managing performance becomes less about navigating your way through who says what, and more about interpreting the data generated by following a process.

Activity measures

A holy grail for performance management has been the quest to measure performance, automatically, without asking anyone a question, and completely on the basis of worker activity. The purported benefits of algorithmic performance measurement is that it is unbiased, and distraction-free, For example Swain et al. (2019) have investigated the use of passively sensed data to characterise the behaviours of top performers. By correlation, this model could be engineered into a performance measurement tool. Likewise Mirjafari et al. (2020) used 'ground truth' employee rating data to analytically derive the qualities of top performing employees, showing that attributes such as the frequency of phone usage was linked to performance. The approach famously used in Amazon's fulfilment centres where the work fits the classic "time-and-motion" model of work and activity is directly connected to output, but knowledge workplaces are also seeking to implement them (e.g. Nikolic).

There are five main reasons, I believe that this approach is flawed:

1. Firstly, models that associate behavior with performance data are built minimising the statistical variance (error) between the overall model and the individual observations. The best models usually have acceptable levels of variance (error) at an individual employee level, meaning that the model is great at explaining the overall patterns of performance, but unreliable at assigning performance to each individual. In practical terms, how much error will organisations be willing to accept from an automated system? Is it ok to be wrong 1%, 5%, 15%, 50% of the time when determining individual performance? I wouldn't want to have that discussion with an unhappy staff member.

2. The second issue is the tendency for intrusive surveillance to creep as algorithm builders seek more data to improve their models with. Just how far down the rabbit-hole will an organisation be willing to go? The discovery that the rate at which people check their phones is negatively related to on-the-job performance might see that data being baked into performance measurement, but the same study (Mirjafari et al. 2020) also found that the correlation only held when the phone was used for personal rather than work purposes. What to do in this situation? The data scientist will have a strong incentive to determine the purpose of the call, perhaps by programmatically eavesdropping on the conversation. It may be a race to the bottom, especially if a business has installed smart-toilets! Yes, pun intended.

3. Surveillance risks aside, as with the Hawthorne effect, if performance is measured on the basis of telemetry data from wearables and other devices, people will adapt to how they work to optimise their metric

rather than achieving the business goal. You may have heard of people attaching their fitness tracker to their pets to increase their stats when trying to impress their friends (I don't know if this works), but its simple to see where the unintended consequences of activity measures might degrade business performance.

4. Passively sensed quantitative data misses out on a lot of information that can only be gleaned qualitatively. As a simple example the high volume of email exchanged between people tells us that they are strongly connected, but it tells us nothing about whether they collaborate closely or are in conflict.

5. Finally, and most importantly, having to participate in a conversation and ask questions can lead to a development discussion, and build a environment in which workers can be supported. Eliminating discussion about topics that impact people is dehumanising, bad for workers, and bad for the organisation.

Activity data is the highest volume, but lowest value data about workers, particularly in the office. If it is to be used, it should be used in conjunction with, and tempered by the results of visible work, and peer feedback. Microsoft's previous addition of a 'productivity' measure in its Office suite was an interesting case in point. It was initially released with individual employee names, so that personal 'productivity' could be seen. In reality, it was a measure expressly geared toward quantifying the adoption of Microsoft's tools. Following community feed(push)back, it was modified to redact individual names and provide aggregate measures instead. Productivity was, and is, a misnomer as it revealed nothing about the quality of work output.

It is easy to understand why it is tempting to jump from simplistic measures such as digital activity to performance measurement. There is no effort required from a manager and no potential for human bias. However, the approach may be counter-productive, rewarding activity over output, and is eminently gameable.

Incremental influence

Bersin (2019) argues that the next phase of employee experience may be having an AI-enabled system that can deliver personalised data to employees to help them manage their own performance. In the context of the Conversational Enterprise this data would be delivered conversationally, by a coaching-bot. Sure these discussions may lack personal empathy, but they may not need too. The fact that are initiated by an algorithm in an automatic and data-driven way makes them seem less 'attacking', while also ensuring that these kind of development moments actually happen.

Additionally, a coaching-bot that keeps track of its suggestions and employee responses, along with peer, manager and activity data is likely to provide a reasonable log of attempts to improve performance.

Coaching-bots offer a new framework in which both managing performance and performance management can be re-imagined.

- **Is it easy to know how effectively people are contributing to work outcomes?**
- **Does your business have any processes intended to help increase performance that are a waste of time? If so why are they retained?**
- **Who is evaluating performance (if anyone) and are they in a position to have a realistic basis for doing so. Who is best placed to provide input on performance?**
- **Are rewards and opportunities given on the basis of good performance, and if so is this just creating a situation of bias and widespread animosity?**

Nudges galore

The "nudge" is subtle method to change human behaviour without overt control. It relies on providing a gentle incentive for people to do something different than they may have ordinarily done. Influencing individuals through the application of digital nudges is already proving to an effective and low-cost method of modifying organisational behaviour. The Conversational Enterprise provides a simple framework in which nudges designed to align an organisation around its intent and enables the achievement of management goals can be easily and intelligently applied, without reverting to human mediated command-and-control practices.

In 2008, Thaler and Sunstein introduced the "Nudge" to the mainstream, a concept drawn from behavioral economics. By shifting the decision making architecture in less overt, non-mandatory, and non-instructive ways, society, consumers, and organisations can be encouraged to change. In the workplace nudges can take many forms (such as the measurement creating a desire to perform, physical placement of objects to change default choices, and subtle reminders that prompt action). Because the remote workforce connects over digital channels and the office **is** the communication platform, the Conversational Enterprise can deliver a range of nudges directly into a workers daily work experience.

In a 2019, I had the opportunity to conduct a live experiment with a client. Employees had previously been canvassed for ideas and issues of work practices to improve, and meetings were targeted as a culprit for inefficiency and frustration. People already knew what was expected in terms of necessary behaviour but they weren't applying this knowledge. A training program had been developed but it was expected that this would likely be ignored (after-all people knew what they were meant to do already). Together we designed a simple nudge that would reinforce this knowledge as well as providing an incentive to apply it. As it is not always easy to 'pull people up on the spot' or 'call out inappropriate behavior', the nudge consisted of people receiving an email message automatically after attending a meeting, requesting feedback to evaluate a certain characteristics of the behavior that was on display (punctuality, listening, etc.). The nudge worked because while the reminder about the expected behavior was explicit, people implicitly knew that their conduct was on display before attending meetings, and would be evaluated. As with the observer effect, the act of measurement changed behavior and yet there were specific elements of behaviour that changed for than others (being prepared was top of the list). This resulted in an overall improvement of 10% in meeting effectiveness, which given the employee time and investment in meetings, was of significant benefit to the organisation.

This example is similar to that described earlier, with team members being surveyed occasionally, asking who they had helped, and who had helped them, with results simply being shown to the group. The act of analysis of a specific aspect of behaviour, and the social pressure of displaying results led to a decrease in conflict and an increase in collaboration behavior (Waber).

In another nudge, a wellbeing balance campaign was being run at BCG, and managers were being encouraged not to put unnecessary pressure on staff with after-hours emails that they were expected to respond to. Rather than leaving it there, and communicating expectations, the nudge was embedded by setting up a trigger as managers clicked 'send' after a certain time. Their email client issued a reminder to the manager about the personal cost for workers of receiving after-hours emails, and a dialog asking whether to proceed (Dhar et al.). It was opt-in, but its resulted in a change in attitude to communicating about work after-hours, and a decrease in messaging related to it.

Likewise, when Google automatically sent an email to managers when new starters commenced, nudging them to take simple steps to onboard new staff by discussing roles and responsibilities and to schedule regular check-ins, workers were found to get up to speed a month quicker than otherwise (Grant, 2019).

With remote work, and with the rise in popularity of nudge theory, we will see an increase in the adoption of the digital nudge. The technology exists to deliver, measure and learn what works, and unlike the kind of algorithmic management that fires workers without human intervention (e.g. Diallo) it can shift behavior little-by-little. It is one thing for workers to be aware of what is expected, but changing behavior with inexpensive nudges is a better way of preserving freedom and trust than implementing an invasive surveillance infrastructure.

Personal awareness

When it comes to data at work, I have argued already that workers should have the right to see the data that organisations hold about them (this subject is now being considered in the courts e.g. Benzinga.) I also believe that this right also imparts the responsibility on the worker to use the data to help them be their best. Rather than heavy handed punitive action based on activity measures, AI is also being used now to equip workers with insight into their behavior to help them increase their effectiveness. Through an automated system of analysis, it is at least conceivable that individual workers will be able to benefit from customised analysis that they are in control of, that potentially nudges better performance and positively incentivising personal development.

- **When your organisation is trying to influence how people go about their work, is its first reaction to tell or to nudge?**
- **Have you implemented any nudges in your digital channels that may be experienced remotely?**
- **The impact of nudges is measurable - do you have the means of measuring them in place?**

Bots can have tough conversations too

Depending on their personality, managers with human emotions can find it hard to have tough but necessary conversations with their staff. One of the strengths of bots is that without emotion, they "find" it easy to initiate and/or take conversations to important places where humans might find uncomfortable.

Managers generally dislike initiating tough conversations as it can push them into being uncomfortable. Lencioni describes this phenomenon as "Conflict Avoidance" in "The Five Dysfunctions of a Team." Equally, managers can also find it difficult to manage conflict between team members impartially, due to the fact that they usually have close relationships with some of their team. It's messy.

The challenge of having productive tough conversations has increased with the move to remote work, with fewer opportunities for person-to-person discussion. With an increased reliance on asynchronous, text-based communications (where it is easier to assume malicious intent due to the decrease in subtlety that can be imparted) there is an increased likelihood of conflict. Necessary conversations are therefore sometimes avoided.

Remote work has also revealed that difficult conversations have been suppressed over time. As an example, the public complaints of discrimination by Apple staff have emerged as a result of organisation-wide discussion on their internal #remote-work-advocacy channel (Schiffer). It has become harder to conduct productive conversations about difficult work topics, and yet with new patterns of work, they are more likely to arise more often.

As bots lack the emotional machinery, but are developing increasingly human-like qualities, their role in conflict resolution may become more prominent over time. Their impassive characteristics may prove to be a natural strength when emotional heat is applied. Algorithmic sentiment detectors can sit in communication channels and interpret messages between people. This makes them well equipped to spot escalating tensions. Anger detection already operated in call-center operations where angry customers are routed to more highly trained staff. Its application within the office context will be relatively straightforward (at least in technical terms).

In time, bots may also be able to directly intervene and help parties resolve disagreement. Bots are already capable of synthesizing complex issues into key ideas, and interpreting the associated sentiment. Algorithms are also competent at interpreting legislation, and using contextual inputs to make accurate determinations and rulings. The basis for conflict resolution is already there. Estonia, a leader when it comes to adopting AI in government, is currently planning on implementing an algorithmic judge for small claims of under €7,000. If successful, this may pave the way to being able to settle workplace disputes in a similar manner. But even without going so far as to settle disputes, bots working in teams may simply be able to bring issues to the surface, and facilitate discussions between people, and help individuals sort out their own problems - something that managers seem to be generally unwilling to do.

From a coaching perspective, it can be a challenge to point out weaknesses or failures in others, but AI-based systems (if configured to do so) will have no qualms in doing so. A bot can consolidate activity and feedback data, and package the results with personal suggestions for individual and team improvement without the fear of personal bias or political agendas. AI agents may make the workplace one in which honesty becomes routine, and conflict avoidance much rarer.

- **Do your managers avoid having the hard conversations with staff?**
- **Can you see a role for bots assisting in conflict avoidance?**

Reducing organisational drag

Sustaining the middle management and executive layers of organisations requires work, the cost of which can be considered a tax on their introduction. By stripping out unnecessary bureaucracy, organisations can release otherwise locked-up productivity.

The workplace is full of distractions that take workers away from their work. Many of these distractions have been necessary to increase alignment and to govern investment and delivery. New ways of working make much of this unnecessary, but they are maintained out of momentum, postponing change and prolonging the state of management, and the roles of managers.

Management check-ins

Catching-up and staying connected is part of the regular workday for most people. Being highly connected is not only a social positive for many; it means that work can be coordinated, identifying opportunities to share whilst avoiding duplication. In the past, work could be planned from the top, decomposed into small tasks, and cascaded through management layers to the front-line. This is how businesses with the 'one-roof-four-wall' business model operated. Adherence to the plan ensured coordination, and the manager's job was to control execution and adjust where necessary to maintain the plan. However, new ways of working that emphasises creativity, innovation, workplace flexibility, and autonomy cannot be managed this way.

Modern work is more bottom-up in nature, reflecting the notion that the executive and the management layers do not have a monopoly on ideas, and cannot be across all the complexity and detail (except in the most exceptional cases). Rather than being imposed, people work at different times, and ideas need room to develop and bounce around in a non-linear fashion. The front-line can also respond immediately to issues at hand - introducing an approval cycle just adds cost. Thankfully modern task management systems permit asynchronous work as well as individual and team autonomy, without sacrificing visibility and coordination. In fact, one of the key benefits resulting from the adoption of work-from-home tech has meant an increase in transparency of the specific tasks that people are working on, and the dependencies between them, in a manner well characterised by Gitlab's remote manifesto.

For the inexperienced manager, motivated by a need to feel 'in control' and a nervousness that they are not, it can be tempting to fill the day with meetings to make sure that everyone (including themselves) are 'in the loop'. This provides them with a sensation that they add value and keep things on track. Recent research shows however, that since the pandemic started, 80% of workers now feel that their immediate manager is unnecessary for their work (Bolden-Barret). Filling calendars with one-on-ones, team chats, all-hands meetings by video can give the illusion of busy-ness, but it is mostly unnecessary. It is an example of "Organisational Drag", which can account for 40% difference in output between those companies that actively seek to reduce it, and those that don't. When compounded over a decade, this difference can be as much as 30 times more output for the effective businesses (Mankins and Garton). This is a huge cost that no business would ever knowingly accept, but because it is an invisible and sunk cost, this busy-ness is rarely challenged. The other consequence of this waste is that it usually doesn't take long before people vote with their feet, and start becoming unavailable, dropping out of meetings for 'technical reasons', but in reality through frustration. During COVID-19 isolation, even my children discovered after a few weeks of learning at home, an increase in the rate at which their classmates started having "connection problems". Some entrepreneurial/naughty kids changed their profile name to things like "Peter - attempting to reconnect," hoping to fool their teacher. People are creative avoiders.

Once task-boards are established, the linkages understood, and people are accustomed to coordinating work with them, the need to check-in on status basically disappears, and discussions can focus instead on other more valuable problem-solving.

Status 'Reporting'

Executives like to ensuring that work is progressing with enough opportunity to correct course if needed. Most workers understand and appreciate this need. However, workers don't understand why they have to report the same information repeatedly and in different formats, especially if they are having meetings as well where verbal updates are provided. This can be interpreted by workers that the management layer is fragmented, and lacks the kind of coordination and cohesion expected of everyone else in the office. Worse, it can be a sign that the management layer's main priority is reporting rather than the working, with little time to get involved in day to day activities. This kind of manager is easily identified. When confronted with specific questions in their area of responsibility, they habitually answer with "I'll get back to you" or "let's take it off-line." They are so absorbed in reporting up that they are not engaging down.

Most businesses have a proliferation of email, wikis, spreadsheets, and task management tools. There is usually no shortage of tech but in many cases, the management have not invested the time to learn how to use them - they are simply too busy to do so. This has the effect that they impose outmoded reporting methods (such as spreadsheets and templates) on those use such tools daily. What's more, with remote work, many managers feel a degree of anxiety about whether staff are working or not, as evidenced by the increase in spyware use referenced earlier. Some managers have started reporting that they feel exposed by approving timesheets without being able to see people doing their tasks. The end result is that data gets updated everywhere, causing duplicated work, numerous versions of the truth, increased organisational drag, lost agility, and a loss of employee morale. The answer to this kind of anxiety is neither spying on staff, and nor is it *more* status reporting. Status reporting doesn't need to be this way, and a management layer that imposes this work to justify its existence isn't needed either. Work that is well managed is self documented and self reported.

In The Conversational Enterprise, bots can access information and status from back-end systems, seek extra information from people when and if needed, and if anyone wants to know the status of work, the answer is just a question away.

Communication

Another huge source of organisational drag is in the creation, filtering, and reading of email and other notifications. Searching for that message that had all the important details in it is something most of us are familiar with. The cost to individuals and organisations is well documented, with workers spending 69% of their workweek on tasks that don't appear in their actual job descriptions. This includes 28% of their time reading and answering messages, 19% searching for and gathering information, and 14% communicating and collaborating internally. Only 39% of the workweek is spent in the role-specific tasks advertised in the job description (Chui, *et al*). I can't imagine anyone applying for the job ad "wanted: someone to read and respond to email, search for documents, and attend meetings" but that is a better approximation time-wise for what is usually actually on offer.

In this context, a conversational AI application can greatly reduce drag, and allow people to focus a lot more than 39% of their time on their work. It can help surface relevant information as needed without the worker performing time consuming search and scan ("where did I see that mentioned?"), it can trawl information repositories to determine whether information has been suitably labelled and stored according to information management practices, and it can read *all* the content and regurgitate the most relevant pieces filtering out the noise, alleviating the need for well designed navigation within information applications. For volatile information, such as that relating to people and processes, organisations are often reliant on key personnel with strong corporate knowledge to act as pointers for others. A well educated coordinator-bot can keep tabs on this changing data.

Unnecessary managers create unnecessary work for everyone else. The Conversational Enterprise can help reduce drag through the automation of many of the "information sharing" tasks that consume so much time, talent and energy.

- **Are people's days becoming increasing full of distractions from work, or are they now able to focus more on work?**
- **Do management activities increase or decrease the burden of work on individuals?**
- **What would be the impact of stopping status 'check-ins' in your workplace?**
- **Do people spend to long finding answers to questions and working out who can help them?**
- **Is communication efficient?**

Business planning

Due to the time, effort and number of people involved, formal business planning happens relatively infrequently. The Conversational Enterprise will help transform planning into a continuous process devoid of spreadsheets, documents and workshops that cascade up and down the business. Rather, it will connect the executive directly with workers by efficiently distilling ideas and suggestions, and communicating decisions.

There are at least two kinds of business plans. Sometimes they are linked and inform one another, but often they are separate.

Plan 1 - The intent

The intended plan of a business contains the decisions about what a business will attempt to do. It usually incorporates financial planning, strategy development, service roadmaps, workforce planning, and infrastructure design. It's the culmination of a juggling act that outlines the goals that have been set, and associated target measures. It balances hopes and aspirations, resourcing and capability, and reflects the current state as a starting point. This plan is often an expression of the will of the executive and it is not intended for wider distribution.

Plan 2 - The digest

This business plan is a document intended for everyone. In it, employees can see how they fit into the bigger picture, and be motivated by the business direction. It is often created by teams outlining what they want to achieve, which are rolled up into branch and department plans, and into an overall corporate plan, which is then massaged into a beautiful document once the rough edges are smoothed out. It builds a sense involvement and generates focus because it is more bottom-up in nature and designed to help individuals move in the same direction as the business.

Integrated planning

Some organisations do only one of these two plans (especially small ones), but most do both to varying degrees. Ideally, these plans inform one another, and through a series of cyclical iterations, a synthesis emerges so that they both say the same thing. Planning becomes a top-down/bottom-up dialog. However, this "conversation" takes a serious amount of to-ing and fro-ing, with managers gluing all the parts together. Because of this, the practical reality is that planning usually only happens once per year.

This can be a problem when an organisation needs to adapt quickly.

Continuous planning

Agile methods emphasise the ability to change direction on the basis of new information. Commonly in software development, new features are planned and versions released on the basis of short-term plans and continuous build, measure and learn cycles - with the goal of baking in adaptation. However it is harder to accomplish rapid cycles of planning at an enterprise level, because of the number of moving parts that are involved, and the overall cost of involving everybody - especially already burdened managers. They are left with a choice of getting involved (and having a voice) or leaving it to others.

To overcome this lag, alternative processes can be implemented to capture short term suggestions between planning cycles. These might for example include smaller business cases for which budget may be re-allocated, mini-updates to parts of the company plan, suggestion boxes, and surveys. These processes might produce changes that fit within the broader plan, or simply depart from it. Rarely is the plan updated.

In the Conversational Enterprise, these dynamics can be brought together into a continuous two-way process. For example, bots can constantly be applying algorithms to conversations and feedback throughout the year to maintain a new kind of 'ideas' register, describing things that are working well or could be improved. The idea-bot can interpret suggestions from employees, distil the main themes and popular ideas, and determine whether issues are novel, emerging or chronic. An idea-bot that is always up for suggestions can maintain an always-current bottom-up list of ideas, and facilitate a process whereby the executive can review and consider this more frequently. No more suggestion that have to be cleared out occasionally, read and discarded. A constant list of inputs allows for more frequent planning cycles which in turn narrows the gap is between the top-down and bottom-up plans, and because they are harmonised on the basis of new data more frequently, they are likely to be more relevant to current circumstances, rendering the stated plan more useful in practical terms. There is nearly no extra cost involved in collecting and interpreting ideas, as algorithms can review nearly an infinite amount of data. Likewise, reminding people about aspects of the plan and tracking its execution can be vastly simplified in the Conversational Enterprise.

The better the business plan, and the more voices that are captured in its formulation, the better an organisation will be placed to tackle its opportunities and challenges. Whereas planning once required detailed involvement from managers to cascade information up and down, a direct conversation between the executive and the front-line can now be facilitated by bots in an efficient manner.

- **Is your organisation able to get useful input into organisational planning from your workforce? If not, what is stopping you?**
- **Would it be beneficial to have an automated process for listening to employees, and continuously synthesising this input for ongoing planning?**

Policy and Compliance Management

Today's workplace operates under a blitzkrieg of new rules, either imposed externally by regulators, or internally as policy and guidance, and most businesses that work remotely will require even more documentation to support staff. In the Conversational Enterprise the task of managing policy implementation is simplified, and many of the policy challenges can be overcomes.

There is a recent trend amongst some to view "policy" as anachronistic, belonging to a management model of a bygone era. Seen as an impediment to organisational performance (and tolerated only when absolutely required) the reliance on policy is seen as implying a talent deficit and cultural shortcomings (for example see Hastings). Policy certainly has some downsides:

1. Policy is expensive to maintain. In the software world, any code that is released and used incurs a technical debt; a liability for maintenance that must be incurred at the expense of doing new things. Once rules are created they impart a similar kind of policy-debt, as they too must be maintained. The standard policy lifecycle generally involves spending weeks canvassing views, drafting, testing, and launching the version. After this time, awareness campaigns and refresher cour-

ses are needed, along with testing and certification regimes. With ongoing feedback, edits are continually made as the policy is maintained.

2. Policy is expensive to understand. Unfortunately, policies are dynamic, and there are man of them. Employees can't reasonably be expected to read, understand, let alone apply them all. The growth in regulation is overwhelming, and banks for example were expected to comply with over 300 million pages of regulation in the EU (Groenfeldt, 2016). The volume is vast, and most executives have no real sense of how many internal policies that an organisation has. If you are an employee how many policies does your organisation have - do you know them all?

3. Compounding the policy challenge is a trend away from prescriptive policy to a more principle-based approach. The benefits of this are that it forces people to genuinely think about how to apply the policy to achieve its objectives in their specific context, and insulates them from the changes that new technologies might otherwise require. However, the approach does impose an extra burden - considering how to implement them. Sometimes people just want to be told exactly what to do.

4. Perhaps the straw that breaks the camel's back is remote work. With staff working anywhere, often at different times, there is an increased need for documentation, policies and written guides (e.g. Gitlab), and a reversion back to favouring formal and written methods of communication over verbal ones.

These factors make the challenge of managing and monitoring policy implementation especially challenging, and there is risk that compliance with policy becomes just another item on everyone's 'to-do' list that surfs near the bottom and never bubbles to the top - at least until the next major breach occurs.

The Conversational Enterprise provides the underlying technology to rethink the place and function of policy, and can greatly simplify a number of the challenges.

1. Effective policy is discoverable. Most workplaces now have a blended mixture of intranets housing some policy content, with pointers to electronic document management systems holding documents, with references back to the intranet where information can change more rapidly, or links to the internet. When its hard to navigate, people ask questions and get answers in their email system. Its a mess. It is also reasonably common to find multiple versions of the same document, with the most recent version having an ambiguous status. Many are thousands of words long, requiring at least skim to determine whether its even relevant to, and can inform some current issue being considered. In contrast, asking a policy-bot that has read all the policy a specific question and getting a suitable answer is far easier.

2. Good policy is easy to understand and addresses real-world situations. Allowing staff to participate in policy management by collecting feedback and suggestions from them helps them pay attention. Testing and validating those inputs with other staff can help refine policy and make it more understandable. This task is difficult for managers to perform, but when information is delivered by a conversational policy-bot, then it is easy to achieve.

3. Policy should keep up with user needs. Where are all those unanswered questions that people have?

When people seek guidance, this creates useful data from which to learn, and this information is not usually collected in a way that allows continuous improvement. It's easy to see what people don't know when they ask policy-bots questions, and to identify knowledge gaps that need filling. Equally, being able to identify those policies and sections that are no longer referenced by people can help identify which should be put on ice, and which might need updating. Policy-bots generate data to help tailor policies to user needs.

4. Policy should be automated when it can be. Written policy and software 'code' have a lot in common, and treating policy as 'code' is a growing trend in the RegTech community. Policy can automated in many situations, especially when they express a set of rules or procedures (rather than principles or expected behaviours). Policy of this kind can operate inside software applications and so compliance with the policy doesn't require employees to understand what the policy is. It just happens invisibly.

5. Where policy is principle-based or behavioural in nature, policy compliance should be measured. This is a qualitative research tasks that surveys have typically been used for. However this creates an extra burden on staff. Why not measure adoption and test understanding as policy is explained by the the policy-bot. This would give a continuous measure of compliance across the business, and allow targeted initiatives to be run where needed.

The growth in policy has created an impossible burden for those managers that are expected to implement it, and the effort involved detracts from other potentially more value-add activities such as supporting and coaching their staff. The Conversational

Enterprise offers a practical means of managing the implementation of policy, by automating the explanation, testing, refining and measurement of its adoption. Already busy managers do not have the bandwidth to manage the growing complexity of policy, but with the assistance of policy-bots, organisations will be well placed Tories to the challenge.

- **Do people understand your policies? Do they apply them?**
- **Is it reasonable to expect that your workers will be across all the details of relevant policies as they change?**
- **How many policies do you have?**
- **Is it easy for workers to discover whether any policies impact what they are planning on doing?**
- **Do you have the right policies, and do they support your actual business needs?**
- **Does your business continue to collect feedback and suggestions about your policies?**
- **Do you have the right policies? How do you know?**

Risk, Incidents, Health and Safety

Automation of various aspects of risk management and incident response can be achieved and delivered throughout the Conversational Enterprise in a simpler, more effective approach than current practices permit.

Over the years, specialist disciplines (such as finance, legal, security, and risk management) become increasingly complex. The modern enterprise ordinarily caters for this by establishing single-function teams, each of which has its own 'body of knowledge', jargon and set of practices. To provide oversight of how the enterprise performs these disciplines, leaders of these functional teams need to rely on those responsible for managing the bulk of the people - the middle managers. Their job requires influence to be exerted over how do their people work. While some influence can come through the administration of systems and processes, other behavioural influence is usually only be achieved via their manager. In turn mid-managers are expected to work closely with them ensuring that those functions are being implemented properly within their own area of responsibility. As the number of these single-function teams grow, the burden on management also increases, and in practice, managers have no real incentive to 'do the right thing' and functional areas are left to their own devices to influence behaviour at arms length.

Risk Management

As a former director of risk management, I understand first-hand the pains of trying to achieve an enterprise view of risk, and to embed its consideration into daily operations. Teams usually had explicit measures, goals and competing priorities that made it hard for them to pay attention to risk (which seemed like a distraction), and so instead treated risk management as a 'tick-and-flick' activity.

The use of AI in the risk context is growing quickly and offers to support the enterprise risk management team in such circumstances. For example, use of passively collected communications data is being used to model organisational risk across the enterprise, without having to achieve buy-in from the manager cohort, and distract people form their work (Gupta and Owusu, 2020). This approach has the potential to a continuously update an enterprise view of risk and show how risk is changing in response to effort to manage it over time. More explicitly, a risk-bot directly engaging with staff to listen to and distil their perceptions of risk promises to alleviate the hurdles involved in organising and conducting enterprise risk reviews.

Incident Response

Incident response is critical to organisational health, and preparation is vital in ensuring that the right actions are taken at critical moments. Depending on the expected incident type, multiple plans, communications paths, and checklists may need to be prepared and executed. As businesses grow in complexity, especially with remote work, the challenge of coordinating incident response becomes exponentially larger. Unless many, many more staff are available to manage this (not likely) business continuity and resilience teams will need to turn to automation, and other data-driven approaches. Technologies that collect and tag incidents, communicate directly with people at scale, automate response actions, and can manage incident response policy will help organisations prepare. Reliance on infrastructure such as that provided in the Conversational Enterprise will also help test readiness under a range of scenarios, proving as much confidence as can be achieved in advance of exercising the capability.

As a worked example, a critical incident might occur in geography in which some team members are located. An incident-on bot is watching electronic channels (such as Twitter) and classifies a tweet as a report, and executes a number of steps that the security team has trained it to do. The immediacy limits the potential damage, and it notifies the security team what it has done, and seeks further instruction. Meanwhile, it messages the affected staff on all channels via SMS, email, chat channels, and the intranet.

A week later, following a post-incident-review the security team has made some changes to the infrastructure that needs to be communicated, but it's not urgent. They let the bot know what needs to be communicated, and because it isn't urgent, it uses everyone's preferred channel, and records the fact that the message has been reviewed when it has. A week following it reminds those staff that did not acknowledge the message previously, and tests that key individuals know the details of the change - giving the security team the confidence that the change has been implemented and effectively communicated. The bot is an extension of the team and works as a communications specialist - something that the security team have limited skills with.

Health, Safety and Wellbeing

In a world of wearables where new devices with new sensors are continually being launched, we can expect the management of workplace health and safety to be seriously disrupted. Wearables are already being used to monitor for physical stress (Fickman), and to guide workers in real-time how to avoid injury

(Soter Analytics). Time will tell whether remote workers will strap on such devices so they can be told to get up and do some exercise (e.g. Cheremond), and whether businesses will be happy enough to have access to health data, but there will be increased opportunity to monitor various aspects of employee health to deliver a win-win. Time will also tell whether regulators are on board too, but the number of examples are growing:

- Algorithms have been proven to be adept in using a mixture of active and passively collected data to predict which employees are most prone to lodge stress claims (Guenolle). By looking at non-wearable data, it was possible to provide managers with an indication of which staff may be under increased stress, and whether extra options should be given to help them alleviate this.
- Data-driven techniques have been used to identify better ways of achieving work life balance and to increase work performance by making uninterrupted time-off a planned requirement (Perlow and Porter).
- The physical environment has been experimented with to help employees get some exercise whilst improving their performance (Ben-Ner). Studying the physical work environment may be particularly relevant for a remote workforce, in which organisations have fewer opportunities to control the physical environment. Wellbeing-bots may be best placed to scan the physical environment to identify hazard using image analysis, and to advise workers on suggested improvements.

Its only a short step to see how measuring stress under pressure for example is something that employers may want to do, and how workers (and their lawyers) might argue that employers had a duty of care to protect workers from situations that caused them mental or physical damage. The fact that some businesses have been caught using video cameras to monitor for employee drug use to protect other employees (Richardson) is proof that ethical use of new types of data will be a matter of perspective, and grounds for dispute.

- **Will algorithms ever be connected to employee health data? Is there a scenario in which employees would welcome this?**
- **Are there any real difference between monitoring on-the-job activities (such as location and movement), and other wearable data such as blood pressure and heart rate?**
- **Is it easy for workers to raise risks in a form that they are visible and handled?**
- **Does your business continue to collect feedback and suggestions about your risks and controls?**
- **Could a bot do a good job of managing the response to an incident? If so what kinds of incidents might they struggle with.**

Employer of choice

Businesses would like often like to consider themselves an 'employer of choice'. In an era where workplace flexibility and work-from-home is the norm, there is a need to provide unique experiences to employees to help differentiate one employer from another. Delivering these new experiences will be the battle-ground for talent.

So you want to be an 'employer of choice!'

Sonya sits in her home office. Happy not to commute, she closes her office door at home so that she can focus, lifts her warm cup of coffee to her lips, and dials into a video conference with her colleagues. Having established the day's tasks, Sonya kicks off her favorite playlist and settles into deep work. She loves this project, and her team is fantastic. Two hours fly by, and the updates to the design are complete. She pushes them through for review. After a quick stroll to clear her head and a few phone calls to connect on other matters, she dials into the next video conference. It's an excellent and productive rhythm, and her workspace fits like a glove. You might think Sonya would say her employer is an "Employer of Choice", but one of the details I left out in this scene was that Sonya has two part-time contracts with different employers. She splits her time between them across mornings and afternoons. She has the tech to be connected, can keep all the arrangements professional, but her premises are hers. Her two different 'workplaces' are physically identical - or put another way, her different employers have identical facilities.

Who does not want to be an employer of choice? It says many positive things about an organisation. Of course, there are those who complain that the annual awards go to those with the best self-promotion, but it's a brilliant recognition to have if the prize is actually a result of supplying a great experience to your workforce. It makes many things easier when there is a market demand to be join - not only for recruiting, but also for retaining talent.

Many factors shape whether businesses are employers of choice. Most evaluations will include measures of culture, development, leadership, flexibility, and, more often than not, physical conditions. As is the case with Sonya, the physical environment will no longer be a lever that her employers can readily pull. The loss of competitive advantage is excellent news for workplaces with a limited budget who can now compete with the big end of town, with their fancy latte machines, exposed brick walls, and slippery dips. Flexible work is now just an expectation. Being able to work from home is a given, and is no longer a strategy to provide a unique employee offer. While many CEO's are working hard to bring their people back to the office (Belvedere), the fact that people will be able to shift from one employer to the next without packing up their desks represents a new dynamic. Those that want to be an employer of choice, will need to provide differentiated experiences to their workers.

- **Is it important to your business that your workers regard your organisation highly?**
- **Has your organisation been using flexibility or the workplace environment are a strategy to attract and retain talent?**
- **Do you have a plan and the means for knowing why people join and stay in the future?**

Differentiated Experiences

There is no such thing as the ideal employee experience.

If you were to stand in my shoes, you might discover that my preferences change hourly, and research shows that most employees want significant degrees of daily freedom (Mercer). Of course, you can cast the solution to that problem as providing maximum flexibility, but flexibility comes at a cost, and some choices need to be made. In "The Social Organisation" the various kinds of work and the styles of communication that each requires is outlined (Ingham). There is no one way of working, and no single way of working is an enduring preferred experience, but rather different types of work have different requirements. Averaging all the expressed needs would please nobody and it is unlikely that you can cater to every preference, so whose will it be?

This challenge highlights an increasingly important question: To what degree will leaders want to cater for workforce's taste and preferences? One strategy would be to ambitiously seek to meet as many of workers' desires as possible; another is to seek employees that want what is already on offer. How this question is approached will have dramatic effects on being an "employer of choice". If a businesses strategy is to be responsive, it needs to understand its workforce. If it is to find employees that suit what is on offer, it needs to understand how to articulate what it has and how to find suitable people. In the past the only options where to run surveys and workshops, and have managers meet with staff and cascade this information up. However, the Conversational Enterprise would see this process occurring as a continuous dialog, with an experience-bot collecting data in the background about employee preferences, and measuring the changes emanating from workplace experiments designed on the basis of that data. This is a far more efficient and continuous way of gleaning insight, as well as measuring the positive (negative) impact of initiatives on employee experience. Bersin (April, 2019) argues that 'employee listening' will become a fundamental capability of organisations into the future, not just to win prizes as employees of choice, but to compete in a more global and competitive talent market.

- **Have you evaluated the impact of remote work on the employee experience? How do you know whether you are providing a compelling employee experience?**
- **To what lengths do you want to cater to the needs and desires of workers?**
- **Does your workplace appeal to any specific groups of workers with different needs, and if so why?**

I'm WFW tomorrow

A physical site for work may provide employees with many benefits, and will probably continue to play an important role in making organisation attractive and productive.

Working From Home (WFH) and Working At Home (WAH) became popular acronyms during 2019 and 2020. WFH has become the norm for legions of office workers, and the acronym shaves weeks off the average worker commute times over the course of a year. During the first few months of widespread working from home, reports of increased productivity and employee satisfaction dominated, bringing into question what the purpose of headquarters were, but as time went on, answers started emerging as people started expressing desires to get back into the office (Seabrook), and leaders started worrying about productivity.

Given the increasing trend to WFH, I wonder whether we will start seeing the emergence of a new acronym. Working From the Office (WFO), or better yet Working from Work (WFW), may begin to appear in people's signature blocks to let everyone know about those rare times that they might be physically available to help schedule face-to-face meetings.

unReal-estate

Naturally, with a shift toward a distributed workforce, we might ask "what happens to the existing real-estate". Some are shrinking the existing footprint, while others are ditching it altogether. Few, if any, are gobbling up more premises. Realistically, we might expect that the nature of a physical workplace will play a different role in days ahead. There is a case for this, and I'm sure that economically the office rental industry will decline for some time without reinvention. Perhaps this is why governments have been so eager to get their public servants back into the big cities, to undo the de-population and draw back the crowds to struggling service businesses.

Why would an organisation bother keeping physical premises? It's a good question, and over time some won't. Some started their businesses without premises (Fried and Hansson, *Rework*, GitLab), and it would be tempting to imagine a world with no offices. However, there are some excellent reasons why physical workplaces will last, especially for larger organisations.

Presence

I have many friends who work in legal and advisory firms, and I'm often surprised by the variation in hourly rates that their businesses charge for what appears to be similar services, though they are all high. Cynically (I hope), a couple of them have told me that the variation comes down to the quality of marble in their respective receptions and they drinks they serve their clients. I wonder how they will adjust to clients that no longer visit, and if the consulting industry will be subject to a shakeup as a result. Businesses have spent years developing a physical presence that communicates who they are to prospective clients as well as their workforce. I'm sure you have friends who have taken jobs because they were so impressed with the offices. One reason to keep a physical office will be to leave an impression of success, much like an expensive business card and a fine watch.

Motivation and wellbeing

A workplace not only supplies a physical location that supports the work, but it also creates a psychological effect on ambition, creativity, professionalism, wellbeing, and teamwork. There is a case that the physical workplace will retain an essential role in facilitating talent and motivating people to be their best. It's hard to do that from the couch in pyjamas. The isolation experienced during the quarantined work conditions during COVID-19 demonstrates this. For many, cabin fever set in while working at home, and the challenge of focusing on work while pets prowled, and spouses put on their 'work voice' on the phone was a significant distraction. In some more fortunate geographies, people were able to safely get back into the office, and to the surprise of some - many they loved it. Some are still waiting while others a resisting calls to return (Belvedere). Part of having a physical location in which people can socialise is for psychological wellbeing and there will be a significant portion of workers that want a workplace to go to.

Physical collaboration

High-bandwidth collaboration and physically colocation helps work that requires high-intensity ideation, play, and brainstorming. Whilst Zoom was initially a favoured go to for conversations, workers have discovered that it is hard to achieve the same results over video, in which fatigue quickly sets in. Creativity and subtle relational work are hard (bit not impossible) to do remotely. It's why horizontal/cross-functional teams (Ingham) generally have a preference for being physically together, and why business folk fly long-haul to close a deal. Also, there is a strong case for relationship building (or establishment) happening physically to supplement any ongoing digital interactions, especially for new staff. Of course, this could happen outside of an office context, but occasionally working physically together will probably buy a lot of online grace and tolerance. Time will tell whether your workers will benefit from having a physical office to work from (if you are measuring it) and how often they should do so. Having somewhere to escape to as a group, even occasionally, will offer opportunities to connect in significant ways.

Digital water cooler

The lack of a physical presence means that there is also a lack of an obvious place for "water cooler" or “corridor” conversations to happen. Much has been made of the need to offer a digital alternative to the opportunities for serendipity that they provided (such as general chat channels or randomised lunch/cafe appointments) as serendipity has been seen as an important driver of a number of positives in the workplace.

Serendipity has been credited with sparking innovation, through the cross-pollination of ideas. This may be true for those fleeting moments of sudden realisation and forming new connections that may help over the longer term, but innovation is a more intentional process requiring ongoing commitment and a supportive culture (e.g. Leonardi). Relying on accidentally bumping into one another seems more like a hope than a strategy. Brainstorming is another area in which the “digital corridor” receives credit as an important catalyst, but like innovation brainstorming is an exercise that can be planned, and there are more appropriate facilities for it than a corridor. In all likelihood good processes, facilities and unplanned moments work together.

The “water cooler” is also credited with supporting the creation of a positive culture, through the creation of social bonds between people. This claim seems to have some grounding (Seppälä and Cameron), though it’s not everyone’s cup of tea (or coffee, or non-caffeinated beverage). Getting to know people socially allows people to relate as humans rather than resources, and is also advocated by those businesses that promote written formal communication over informal channels (e.g. GitLab’s Remote

Manifesto). However it is hard to organise spontaneity - it can seem like forced fun.

Where the water-cooler seems to come into its own league is as an informal communication device, to compensate for awareness gaps that grow over time. Remote team communication relies nearly entirely on an explicit model of communication, where people communicate only when they have a reason to, and when that reason outweighs the cost of communicating. In contrast co-located teams rely more on implicit and opportunistic communication patterns which have been shown to increase team, organisational, social, group and context awareness, all of which have been shown to improve planning (being on the same page), task execution, and increase in cultural "friendliness" rather than "exerting authority" (Röcker, 2012).

In diverse workplaces, the kind of informal communication that comes with a digital water cooler also promotes the opportunities for those that hold minority positions to connect with one-another (its nice to know you are not alone). This kind of support can promote diversity, but where an organisation prefers control over information flow, and desires to maintain a culture of secrecy, this can be a threat. Corridor conversations allow for conversations to occur in hushed tones but their substitution with authorised channels can be intimidatingly public, on the record, and subject to sanction. For example Apple's #remote-work-advocacy channel was useful for promoting a discussion between people that didn't regularly work together, but shared an interest in the subject. The connections formed led to comparisons and the identification of mis-treatment by managers, which went unaddressed and led to public disputes (Schiffer). There is a cost benefit to consider, and some businesses might be celebrating the fact that "water cooler conversations" no longer happen.

There can be no denying that the popularity of the notion of the digital water cooler at least suggests a real need. If nothing else, it provides a back-stop to compensate for poor communication. Given that 91% of employees consider their managers lack good communication skills (Schwantes), its probably a necessary one. How to implement one is far from settled though and may be a function of the preferred culture of a business. The Conversational Enterprise is a business architecture that promotes the use of communication channels to experiment with and explore what works, without having to create a one-sized-fits-all approach. This may include of course, just leaving it to staff to just work it out themselves, but could also include a bot-facilitated discussion about topics of interest, where participants anonymity is preserved.

Future footprints

Workplaces are trending toward smaller footprint, but more interesting workplaces - without the vast rows of pods or the hot desk seats of cubicle-farms. The trend toward smaller or shared headquarters, or an increase in the use of satellite facilities with more targeted and branded employee experiences is likely to grow (Seabrook). The fact that hotels are vying to provide dedicated and well serviced executive rooms while demand from the business travel sector is in decline seems to point this way too. I expect that the workplaces of the future will diversify, catering to different worker tastes and look more like playgrounds, but for adults, and the Conversational Enterprise will assist businesses cater for the needs of workers, whilst maintaining connection between them.

- **If your office was permanently unavailable what would be the impact on your people?**
- **Are there things you would do to compensate for its absence? Are you already doing these things?**

Hiring teams rather than individuals

In a world where teams operate more autonomously, why not hire teams rather than individuals? Processes are not currently geared toward this end, but hiring whole pre-existing teams is likely to provide numerous benefits.

There is no denying the popularity of the Avengers movies. Thor, Hulk, Tony Stark, and the rest have had independent careers and periods of individual work along with their own movies. But under the leadership of Colonel Nicholas Joseph Fury Sr., they are a team (that can sell movie tickets) like no other. I'm not sure what the process of getting the Avengers on board entails, but they often come as a set. It took a while for them to become a cohesive force, and personnel changes have happened over time, but like most teams, they know one another's strengths and weaknesses, and together have a unique blend of talents that can help get the job done.

I've often wondered why in the workplace we rarely hire teams. Except for exceptional cases and consulting gigs, it is as common as hen's teeth. The distributed organisation is a great scenario for its wider adoption.

- Firstly, distributed teamwork takes a little while to build harmony. Most teams have idiosyncratic issues that are easy to trip over if you are unaware of them, and these unique characteristics need to be uncovered team-by-team. Hiring a ready-formed team may reduce the time to productivity, and reduce the risk of inter-personal issues jeopardizing the chances of their success.

- Secondly, there is a trend amongst companies to acknowledge that teams are at the centre of how things work. They have started implementing performance management at this level, and have devolved a lot of decision making there as well. With this in mind we are likely to find emphasis on team-based incentives. By embedding reward structures for groups rather than individuals, teams become at least conceptually self-policing because the members succeed or fail together. This is a simpler approach, and is also likely to be more effective than assessing individual performance, especially if the team has a manager that is not deeply engaged with the team, or across the range of skills that the teams possesses.
- Thirdly, teams know what they are capable of, their strengths and weaknesses, and how to divide tasks amongst themselves. Writing job descriptions can be complicated, and even knowing what skills are necessary can sometimes be a challenge, especially for new capabilities. Why not let a team explain together how it can meet the needs.
- Finally, being part of an established team that you have chosen to be part of could be an excellent employee experience. Being bounced from one set of strangers to another is known to create high levels of stress, which can diminish individual performance.

There is currently a lack of infrastructure for engaging teams but a handful of businesses are now focused on supporting this model. For the time being, job boards, recruiters, and the like will still be geared toward hiring individuals, and the technology used by HR teams (such as applicant tracking systems and performance management workflows) mirror that. However, in the Conversational Enterprise, connecting teams to the rest of the organisation's people, processes, strategy, policies and values is straightforward. Hiring individuals often leads to organisational complexity (), but the Conversational Enterprise can grow without adding such complexity as there is no need to consider spans of control and reporting lines.

- **If you had the opportunity to do so, would you hire pre-existing teams, and what qualities would you look for in them?**
- **How do you think team-based incentives would change performance management practices?**
- **Do you know where you would go to source a team to help you?**

Who is actually doing the work?

The Conversation Enterprise will see bots and humans mingling, and sometimes it will be difficult to differentiate one from the other. Should people be rewarded for the contribution of their bots, what happens when we accidentally hire a bot instead of a person, and how should we respond if someone falls in love with them?

Having a bot that can work on my behalf is a luxury. When someone asks a question about a process that I'm responsible for, a bot can step in and I can avoid the distraction. The bot can even follow-up and check that the person asking is satisfied, or whether there is anything else they need. All I need to do is prime the bot with a few key things and I can get work done around the clock. Assuming that my bot works well, I might expect that I am praised by my colleagues, and if I train my bot well, then why not?

Multi-threaded employment

Early in my career, I was shocked to find that more than one colleague of mine ran an online store from work, on work's time. This was explicitly prohibited under the terms of their contracts, and was treated as a severe breach. Since the uptick in remote working, and more flexible work, there has been a growing trend reported for people to run side-gigs outside of work. Dolly Parton's "5 to 9" DIY website advertisement is a showcase of the popularity of the side-hustle. In two organisations I know of, workers confronted by evidence that they were operating another business during business hours didn't believe that there was anything wrong with what they were doing, as it didn't limit their capacity to perform their requested duties. There are few controls for restricting this sort of thing, and even with surveillance in place, employees will be able to work around it. But there are other ways to ensure that people are devoting their full effort to the job that they are being paid to do. Here are some ideas:

- Employees need reminding of their contractual obligations. If that requires that no other employment be entered into, or work of a specific nature or during certain hours performed, then this may need to be reinforced from time to time. In the past it's been fairly safe that most people understood this, but work-from-home is changing its obviousness. Businesses could start by gauging the degree of understanding of these requirements from staff. In the Conversational Enterprise, bots could measure understanding and remind people of their obligations.
- A significant percentage of employees with an unauthorised side-hustle have reported that they were bored, had no clear direction, and could not get time with their managers to discuss the issue. For them, this meant that their work wasn't being impacted. The obligation to support workers is reciprocal, even if only implicitly so. If remote workers are not being led or tasked, and have had virtually no contact with their supervisors, we shouldn't be surprised that they find other ways to stay busy. Are their managers clarifying what work needs to be done, updating role descriptions as needed, inspecting progress, and managing performance? In the Conversational Enterprise employees would have a chance to highlight this issue.

- Knowing that other employees are working hard can be motivating. Making work visible to all team members and being able to see progress (and where work is being held up) can nudge productivity. By using task management systems to allocate work and track progress, its easy to set up check-points that involve others to inspect progress (even as a reviewer only). In the Conversational Enterprise, if individuals are putting effort into a side-hustle at the expense of work, it would be obvious to those around them.

Having disengaged managers communicates that what employees do with their time doesn't matter. If your organisation is experiencing challenges with this kind of side-hustle multi-tasking I'll wager that the managers aren't managing. Rather than setting up surveillance cameras, screen scrapers, and motion detectors, the Conversational Enterprise is a mechanism to hear from staff about their managers, remind people of their obligations, and to track individual contributions to work.

Voight-Kampff Test

In 1968 a work of science fiction was published that foretold of a future 2019 filled with bots or replicants as they were called (Dick). It sounded so futuristic - pure fantasy. Further popularised by its movie adaptation (Blade Runner), replicants walked the streets and were virtually indistinguishable from humans. While one of the governing laws was that replicants were forbidden from hiding their identities. Some did, and the LAPD was there to find these "runners." To distinguish between bots and humans, members of the force administered the Voight-Kampff test - an updated version of the Turing Test. It involved a series of questions that only humans could elicit the right response to (though the movie's plot rested on challenging this proposition).

What if in the workforce of the future, you needed to administer this test? If you are hiring remotely, in the world of deepfakes, generative content, and multi-sensory systems, is it possible you might make a job offer to a replicant? And more importantly, if so would you care? I am waiting for the first news article in which this happens. I think it will happen prior to 2025. As the cost of generating deepfakes decreases, an engineer producing a high quality replicant may be able to hold down many jobs at once. These 'imposters' may lack specific skills, but they may actually get by in certain roles and where managers are not so on the ball. Maybe we will need a workplace Voight-Kampff Test sooner than we think.

- **Have any of your staff taken on work outside of their duties since working remotely?**
- **Do you believe that workers in your business are actually engaged in meaningful work, or are they filling time in return for pay?**
- **Are the managers at your work aware of the reality of what their people do?**
- **If you accidentally hired a productive bot instead of a person, what that bother you? Why?**
- **What would be the impact in the workplace if someone grew emotionally attached to a bot?**

Changing the culture

One of the concerns that business leaders have expressed with the rise in remote work is that it jeopardises the opportunity to build a positive work culture. Middle management have been in a position to reinforce good and bad aspects of culture, but their opportunity to do so is now limited. The Conversational Enterprise will allow for the executive to continuously monitor organisational culture, define the desired state, to hear directly from their workforce, and embed a new culture.

Choose any research into organisational performance, and you'll discover that it will be claimed that culture has a significant bearing on it. Culture establishes the operating parameters for people, and in times of uncertainty and stress, it's the difference between why some companies fail and others succeed. Culture also communicates expectations and behaviours in a way that cannot be prescribed or enforced by policies or instructions, and so it can also simplify the governance of an organisation. The quest for a positive culture is top of mind for many of today's executives, and no one will receive a prize for uniqueness by trying to create it. While it's not easy to shift a culture, there is no shortage of attempts to do so. “Culture eats strategy for breakfast” is an idiom that pervades management consulting, even if its origin is disputed. And culture is not static. It’s a function of executive behaviour and aspiration, and the habits and reflexes developed by the teams and individuals in it. Internal and external forces shape it. Culture develops whether it is intentionally directed or not, and it cannot be put on pause. This reality is being played out publicly inside Apple at the moment, as managers seek to control information flow whilst employees wrestle control of what kind of values the business stands for (Schiffer).

There are some common approaches to trying to change the culture of a business. Most offices I visit (and intranets I see) have 'posters' on the walls describing their culture, and a glossy brochure containing some form of culture statement. It is often a collection of words covering the desired behaviours of the people as a collective. Sometimes those words, as though springing forth from the 1990s, form a neat acrostic. I don't mind, but I find this strangely common, and I haven't been able to work out its origins. The basic idea is that by teaching everyone the acrostic/brochure/poster that behaviour will follow. Also strangely common is the speed at which new CEOs decide they need to do a culture reset after they take hold of the reigns. This often happens when the new boss arrives, and they try to roll out what they perceive as the "ideal" organisational culture.

You have probably seen these efforts, and may have even been tasked with replacing last year's acrostic with a new set of words. An office I recently visited had several layers of culture statements pinned outside the executive office suites, layered and well preserved as if set up to support some form of archeological research down the track. It was sad that the culture statement could be replaced so simply with a thumb tack. No doubt, this activity is all well-intentioned, but the same tired formula is repeated *ad nauseam*, it doesn't work, and it can amount to little more than window dressing.

Remote work

After several months of remote work, the anecdotal data indicates that work has become less 'collaborative' between teams by a margin of approximately 15% (Arkcoll). However, this finding is based on messaging traffic and may just reflect a greater focus within teams, and less inter-team interference. My personal analysis of a small number of businesses in this time does however align with this finding, with qualitative analysis indicating that workers are feeling more isolated from other teams than they used to on a day-to-day operational business perspective. In contrast, new social channels provided to employees means that greater connections are being built across enterprises - not so much for day-to-day work but around employee experience and the voicing of grievances and complaints (e.g. Schiffer).

It has also been suggested that the increased flexibility benefitting workers has led to an improved tolerance of the preferences, approaches and behaviour of others. Perhaps people have become more patient? However, some CEOs have reported contrary views, that their sub-cultures becoming more deeply rooted and isolated, leading to pockets of stress, tension, and conflict.

It will take time for the metaphorical dust of this radical change to settle. This means that, intentional culture change initiatives will remain a fixture in business, but because of new ways of working, their focus and the methods will probably need to change.

A new approach

In the Conversational Enterprise, bots will augment and sometimes replace managers, and will execute campaigns designed to influence culture directly on behalf of the executive. Table one offers a comparison of some of the activities that may form a part of cultural change initiatives.

A fairytale

TRADITIONAL APPROACH	CONVERSATIONAL ENTERPRISE
Executive develop a culture statement, either by themselves or with a small team of insiders and consultants	All staff contribute to the development of a culture statement. Bots, continuously collect information about what staff perceive as strengths and weakness of the culture, and creates suggested improvements for consideration by executive
Posters and brochures are printed, blogs created, desktop wallpapers are regenerated, town hall meetings held. Some people pay close attention.	Reminders and snippets of cultural aspirations are injected into relevant conversations by bots at suitable times to nudge behavior. This messaging cuts through more.
Monthly or annual surveys collect data on culture to measure the delta between current and target state,	Continuous feedback is collected within the work process itself through employee listening/People Analytics techniques.
Management tactics are developed and rolled out to close the gap in areas that require it. Consultants are hired, and training programs put in place to change mindsets.	Campaigns are tested and run by bots on a continual basis. Experiments are conducted to determine what narrows the delta before rolling them out more broadly.

Table 1 'Traditional' and Conversational-AI approaches to cultural transformation

Let's play out a hypothetical scenario to see what a new approach might look like in practice.

Ms. Black is appointed the new CEO of Goober Corporation, which is in financial decline. The Board has formed the opinion that Goober's culture is stagnant, and Ms. Black is the right person to shake things up. When Ms. Black commences work, she gets the leadership team together to discuss culture. Everyone has an opinion, and some have more than one. Ms. Black requests that the team find out from staff what their views are. Mr. Brown takes on this task and rather than running a survey (which will be impacted by the fact that everyone is nervous because of market sentiment and the arrival of a new CEO).

Mr. Brown turns to the data provided by a long-running campaign that has been operating to embed the then organisation's culture statement. For 12 months, a culture-bot has been looking at the organisation, collecting snippets of feedback based upon who is connecting with who, and working out alignment with the various teams' culture statements. The data shows that 'trust' has been diminishing between the corporate and sales areas. The bot has synthesised anonymous feedback showing a common thread that processes are lacking, leading to decreased service quality between Corporate and Sales teams. The lack of performance is hurting everyone, cascading into poor service delivery to customers, and has resulted in finger-pointing. There is no shortage of sentiment that everyone is trying their best.

Mr. Brown decides to share this data with the executive group. Not for the first time, everyone has an opinion, and some have more than one. The leader of Sales thinks that Corporate IT systems need an overhaul, and the leader of Corporate believes that sales people need more accurate revenue forecasting skills. The leader of Culture wants to bring in a consultancy to run a leadership development program over the next 12 months.

Dr. Green, who looks after Digital suggests an experiment, why not run a campaign and use a bot to take care of some of the hand-offs between Corporate and Sales, explain policies and procedures, automate some workflows and notifications, and collect feedback from staff while it is at it.

Ms. Black concurs because it ticks some boxes and is a simple way to start.

Over the month, Sales are getting immediate answers to queries, and the service-bot simplifies their experience. Because of this, Corporate is getting better visibility if what is coming. The longitudinal measure of behaviour alleging with corporate values starts to tick up, and trust is on the improve.

Ms. Black shows the Board what they have done, and justified it with an empirical basis. The market cheers - roll credits.

While simplistic, this fairytale shows how coupling data analysis, scaling it with tech, and some automation can tangibly influence culture.

Revamping the Departure Lounge

Leaving an organisation (whether voluntarily or not) is a big step, not just for the individual, but also for those left behind. It's easy to spot an empty chair in a physical environment, and it's noticeable when Harold hasn't been in the common areas getting his cup of tea for a while. People bump into each other in these spaces (as with the watercolour discussion earlier in this book), and unplanned discussions happen. When we don't see one-another, it's easy to be unaware of someone's departure. When this happens, there is a heightened risk that things might fall through the cracks during handover. Knowing someone is leaving is usually a good trigger to get organised for it. Also learning who the replacement is (if there is one) can be hard if you are outside the team. There are options galore when it comes to digital channels, but thinking about some of those described in this book, maybe a bot with Harold's knowledge is left behind in his place (like an intelligent auto-responder) for some time. The vital information that he knew would be nicely encapsulated this way, and rather than getting an email bounce, it would be more productive for a reply to contains a suitable response from Harold and a pointer to his replacement.

When it comes to staff departures, two related adages come to mind:

- *Your culture is who you hire, fire and promote*
- *Your culture is what you tolerate/walk past*

In the past it may have been preferable to the basis behind keep involuntary departures to a hush, but because of the heightened need to communicate staff changes with remote work, maybe this requires a rethink.

- **Is maintaining a certain culture a useful and realistic goal for a distributed workforce?**
- **Is your aspirational culture a good fit for your mission, and your people? How do you know if that is the case?**
- **What can be done to shift the culture in a positive direction, and how would anyone know if those attempts were working?**

Diversity and Inclusion

During last decade, there has been a growth in the adoption of programs that promote and the needs for and the benefits of workplace diversity. Key to this has been the publication of metrics that reflect decision making at key stages of the employee lifecycle reflect inclusion (or lack thereof). The Conversational Enterprise can support and extend these initiatives through deeper analysis and influence of workplace behaviour.

"What gets measured gets managed," and the decision making that influences the composition of organisations is now widely being measured. With the goal of having workplaces that better reflect the society that they are part of, the workforce mix at key employee lifecycle moments including job applications, hiring decisions, pay, promotions, grievances, development opportunities etc. can indicate imbalances, and be used to track changes over time. The publication of the metrics is a key strategy that underpins this shift, and is seen as crucial for business to obtain the benefits of a diverse workplace.

The Conversational Enterprise provides a framework to not only simplify the reporting processes (by collecting, synthesising and reporting on this data), but also to push the analysis deeper, and to gauge inclusion at an operational day-to-day work level. In a 2019 research activity with a large advisory firm, I was engaged to explore the linkage between gender diversity and work performance. "On-the-job" performance ratings (1 to 5 scale) from the previous period were linked with the communication patterns (derived from email) of staff in specific work-roles, and the gender of the sender and recipient. This enabled every message to be classified as gender-crossing or gender-aligning within gender groups. Assuming that people communicated without bias, one would expect that males and females would send a volume of messages to people in their office that was proportional to the background gender mix of the organisation. A clear pattern emerged. At the lower levels of the business (graduate and junior roles) there was a much stronger tendency for individuals to communicate within their gender group. This led to the hypothesis that new staff have a tendency to rely on a "friendship-groups" in their early career stages. However, we also discovered that the highest performing graduates had a higher likelihood of being "gender-blind" when it came to their communication patterns. Ambitious graduates may be well advised to adjust their behaviour. With increasing seniority, the bias decreased at a steady rate until there was no overall discernible difference between communication patterns and gender balance (of the organisation or specific team) at five levels up the organisation hierarchy. Like high achieving graduates, leaders were on the whole "gender-blind". Of most interest overall was the finding that certain leaders/partners displayed a high bias toward communication (including the timeliness of response) with people of their own gender, a finding also negatively correlated with performance, consistent with gender-bias. This finding provided a basis for follow-up analysis, but having established the means of measuring gender bias in communications, it's a small step to establish ongoing monitoring and reporting of these metrics, and also to extend the reporting to include other groups of interest.

Within the context of a Conversational Enterprise, such data could be used by a bot to automatically trigger reminders on the need and benefits for being inclusive in the office, and to track any resultant changes over time. Such an act would not need to imply that gender discrimination was taking place at any team or personal level, but by coaching graduates, leaders (and others) about the benefits of being inclusive through a series of automated nudges, changes in workplace behaviour that can be linked to performance could be achieved. This could augment current approaches at setting diversity and inclusion targets and reporting against them by operationalising the analysis.

- **Should diversity and inclusion reporting include day-to-day operational activities, or just major employee lifecycle moments?**

Evolving the structure

Changing the structure of an organisation can be a necessary undertaking but depending on how radical it is, it can also be very disruptive. Despite this, organisational restructures are often entered into without the necessary data ground-work to define what specific problem is being addressed, or to measure the effectiveness of the change through implementation.

Reporting lines matter because they determine the allocation and control of resources. Organisation structure has some obvious impacts on how they operate, and numerous studies have focussed on the relative pros and cons of hierarchical, functional, cross-functional, and matrix structures on modern work (Ingham). The impact of no and low-structure organisations has also been the subject of research (Hamel, holacracy.org). The interest may perhaps be explained by the appeal of being able to make one change that addresses manifold problems. Reporting lines are something that can be easily controlled. However, in many businesses, reporting lines play second fiddle in importance to geographic and physical dynamics. Timezones and locations can easily create schisms. Asymmetry of power between locations can generate friction and perceived threats, and personal relationships affect access to political power. Organisation design is complicated.

Popular models

A popular approach with many proponents in the modern workforce is a "service teams" models. Classic examples include Spotify (Kniberg), and the militarily inspired Team-of-Teams (McChrystal). ING Bank (Schotkamp and Danoesastro) have extended this thinking by also placing corporate support resources (such as HR members) within service teams. The key to them is to make the team a self-managing unit, with full ownership of the delivery of a particular service consumed by a customer (internal or external). Doing so, it is hoped, minimises inter-team hand-offs, simplifies decision making, and concentrates communication within teams, thus freeing the team to explore and solve problems. Because of their relative independence, achieving alignment across these teams can become problematic. To address the risk that separate teams fail to deliver cohesive organisational outcomes, several frameworks have been advocated (e.g. Scaled Agile).

Lacking data

Unfortunately, re-organisations often kick off spasmodically without proper consideration of the risk of a poor result, as though the most challenging task involved is drawing the chart and making the announcement. Even those implementing the more trendy models often do so without the kind of agile 'test-and-learn' approaches that the models themselves encourage. In the absence of evidence, the design is usually worked out by the management in the back-room with some whiteboard markers and post-it notes, based on intuition, and jostling for power. Sometimes the design is made without an appreciation of what the real work that their teams do is. Most of the time, the plan is a balance of factors relating to team sizes, pre-existing seniority levels of team members, personal relationships, and to issue rewards and punishment. In the worst case, a new structure emerges from seemingly nowhere that satisfies and rewards a few, annoys most, and reinforces or creates more problems than it solves for everyone else. And because such designs emerge in the absence of data, there is no possibility of tracking design effectiveness, meaning that you only get one shot at doing it right - high risk.

The natural organisation

Fortunately, one factor stabilising this risk is that a natural organisation operates behind the designed one, and life may go on as though the leadership had never announced the change. Of course this can also inhibit change if one is unaware of the shape of the natural organisation. People often find communication pathways to get things done no matter what the structure is, and organisations can use this fact to their advantage.

Network analysis

The mechanisms for describing the 'natural organisation' has been a fertile area of research for decades, and options for doing so are now widely available. By discovering the nature of relationships between people, using quantitative data drawn from messaging and collaboration tools, or qualitative data sourced through surveys and interviews, it is possible to visualise how work naturally happens. That data is valuable, and can help propose a structure that minimises friction between it and the natural organisation. By reducing the number of organisational boundaries that people have to cross to get work done, the number of necessary hand-offs and approvals can be minimised (or so the theory goes). It is a formal approach to 'getting out of the way' and making the natural organisation explicit in its formal structure.

The software engineers reading this currently have lights flashing in the part of their brain labeled "Conway's Law". This Law goes back to the 1950s and is one of the more enduring principles of software engineering. It says that any system that is produced will reflect the communications structure of the organisation that created it. When applied in reverse, the Law might be used to design a communications (and reporting) structure that better delivers an organisation's intended services, with teams acting like micro -services.

However, care should be taken when applying this kind of thinking based on data. The designed organisation will still have some presence in the network analysis and any data collected will have some echoes of the 'as-is' state in it. People may appear 'central' to work and highly connected, not because they are vital, but because they are a bottle-neck and most quantitative approaches to network analysis do not characterise the nature of the relationship in terms of positive or negative, just their strength.

Monitoring and adapting

Relationship data cannot always be relied upon to develop an organisational structure, but coupled with other data including process deficiencies or lags, communication quality, the availability of rare skills and resources, network analysis may be of significant assistance. If the goal is to shake things up and give people more of less influence, such data is unnecessary. But once data is applied to inform the design, the impact of the change can also be monitored over time using it by repeating the analysis, and adjustments made based on the observations.

The employee experience of restructuring can be one of confusion and dislocation. In the Conversational Enterprise, this experience can be improved because of the ongoing dialog it can provide, the approach to testing change, and because of its data-driven nature, the rationale for the changes can be provided to employees, thus alleviating concerns that decisions are being made for political reasons.

- **What has motivated any restructure that you have been part of or initiated in the past?**
- **Did it get the results that were being sought?**
- **Do you know of any methods by which organisation structure can be designed that link it with desired business outcomes?**

PREPARING FOR THE FUTURE

As the Conversational Enterprise and elements of it rise in prominence, there are benefits and risks for businesses and our society to navigate. In section nine I encourage readers to be active in helping steer the direction of workplaces toward a positive future no matter what their role in it, but just as with oxygen masks on aircraft in an emergency, to prepare themselves first.

Preparing yourself

In light of the current and impending changes that bots, algorithms and and new work practices are bringing, you have a range of personal options to consider in how to respond. The good news is that you probably have five to ten years before the real impacts are felt, and so if you start preparing soon then you will have plenty of time.

You may be an executive leader thinking through the future architecture of work, a manager considering your role in it, or an employee wondering how you will relate to your colleagues and managers in the future. Whichever the case, it would be natural to be thinking about how you will respond to the changes that are heading toward us.

Just remotely managing

Never waste a crisis, they say, and the COVID-19 pandemic has been a doozy. Amongst the many things it has accelerated, the trend to remote work went into immediate overdrive. Moving swathes of office workers have caused CEOs to tweak and upturn operating and business models, and in many cases - it has worked. Having participated in this experiment, you will know that the future of work will continue to change. We won't be snapping back to where we were. The change is permanent, people have altered courses, and the attitude toward future change is more open than it previously was.

Amongst the more critical realisations of this period is that a greater focus on maximising the value from a businesses' highest cost, its talent, is needed. It became simultaneously more complicated and more important to do so. Processes that used to require physical contact can no longer be used, and simple digital analogs don't work. Without being able to digitally 'walk the floor' managers scrambled for information about what was going on. Spyware took off, and best practice guides were gobbled up. This helped with the initial survival phase, but the jolt also got people thinking about management practices more broadly - perhaps for the first time in decades.

Combining a growth hacking mindset, new data analytics, agile practices, and new means of connecting with workers, the Conversational Enterprise provides organisations with an approach that will enable them to continue to adapt in a data-driven way. Those businesses that succeed will continue to experiment, learn and reinvent. Embracing these new capabilities will allow business leaders to do far more than just remotely managing.

Embracing the Conversational Enterprise is not without risk, and leadership will be needed to steer our professional lives away from a potential dystopian Big Brother future. This requires that leaders are familiar with AI's capabilities, surveillance, and the underlying issues associated with data, bots and ethics. If you are in the privileged position of having choices about how data is used, I want to encourage you to develop more than a passing familiarity with these issues, and to build the kind of business that you'd want your children to work in.

You may also have a number of middle managers and workers in your business that may become concerned for their future. For them the personal implications are probably more significant than they are for you. Perhaps you can help them develop a sensible plan for their future.

Return to your expertise

For managers that have risen from a more technical or delivery role who are concerned at the shrinking role and need of middle management, one pathway forward is to return to your expertise. One of the challenges associated with this approach is that the required knowledge may have moved on and you might feel that you may no longer possess the skills, expertise, or certifications to be competitive. You might be right, but the good news for you is that you are probably over-estimating the size of that gap, and what is considered to be proficient these days is generally a lower bar than when you were working in similar roles.

The internet has a lot to answer for. I have recently spent a stint learning some new technology, and the number of examples and instructionals available was incredible. It made making a start and becoming proficient 'enough' to work through the details quite straightforward. For anyone with a decent grounding in a subject, you might just be a few YouTube videos away from being able to start learning on the job. In any case, the rate of change in many fields and especially technical is so quick. The half-life of expertise has shrunk and so everyone is in the same boat. Workers now need to constantly up-skill, and survival depends more on adaptability than on tenure or experience (Ceremony). Returning to your expertise and the pursuit of technical excellence can be a rewarding path, especially if you have not been passionate about your management duties and are a little jaded by them. Maybe being a team leader would give you a chance to blend this expertise with looking after a few people - something you've had a chance to practice.

Invest in your humanity

Bots and people are different, and human skills have an enormous role in the workplaces of the future. However, many managers have struggled with doing human things well, and few have had formal training in aspects of effective management and leadership. One approach to seeing the advances of AI eating into the space occupied by managers is to focus on what makes you different, what machines will never be able to replace (your charm, wit, inclination toward servant leadership, your emotional dexterity, interpersonal influence....) and will make you out-complete bots. This might be a significant leap, but it may also provide substantial longevity and satisfaction. Good managers are rare, after all (Randall and Harter).

Learn some AI

A third approach advocated by some (especially training schools) is to invest in learning about AI. I wouldn't suggest that learning something is harmful, but I have encountered some at the cross-roads considering getting a dose of AI knowledge under their belts and onto their CVs. New courses are emerging every day. Some run for just a few weeks, and claim to give you everything you need to know to be a data scientist. The reality is that becoming skilled in AI is not a short course, nor is everyone cut out for it. People that understand AI have usually spent a few years in postgraduate research, and a short course is unlikely to help you contribute meaningfully to that work. I'd offer two further personal perspectives that might support this view, though admittedly I think I'm in the minority.

1. The tools needed to get involved in AI are getting simpler every day. Professional data scientists now often work with systems with approachable user interfaces, and AI systems are being productised to such a degree that you can make practical use of them without knowing the nuts-and-bolts of how the algorithms work. For example, the various cloud providers have toolkits of pre-canned AI ser-

vices that can be chained together relatively easily to deliver solutions. It will get easier still.

2. The role of a regular data scientist found in the businesses that consume these services is reasonably automatable. Machines will be training machines in no time at all for a large chunk of work that today passes as analytics, data science, or machine learning. Being somewhat skilled in this work without being excellent is likely to be a short career path.

By all means learn where AI fits, its strengths and weaknesses but the chance of suddenly switching to being an AI expert are low.

Ride it out

Of course, there is always the option of just riding out the change and taking your chances. There are think-tanks focused on making life easier for you, and the government is thinking about providing a pathway into the future for you. With a little bit of luck, they'll do a good job, and with a bit more luck, all of this will blow over.

Learn your rights

As an employee you have legal protections. The emergence of algorithmic management has led to many to feel as though their rights have been eroded, but regulators are starting to pay attention.

In law, consumers are gaining strong protections in how their data is managed and used. But laws don't provide any protection unless someone asks for them to be enforced. In this book I have outlined that some businesses will be very careful in how they use data, while others will try take every advantage that they can. An educated community of workers that understand their rights, and can lobby for them to improve is one that can be protected.

You will discover that the popular media covers these sorts of issues reasonably frequently. When you find a journalist that covers such cases, follow them!

Prepare to meet your synthetic colleagues

Most people are familiar with bots, but few are familiar with competent ones - the kind that can play a role beyond basic task automation and hold banal chats. Keep an eye on how the marketing community make use of bots, and how advances in AI are pushing the boundaries of the human/machine interface.

The development of sensors and wearables will give you pointers regarding the kinds of data that will become available to bots. Imagine that your co-worker bots had access to such data. Keep an eye on the creative elements of AI and see what machines are able to produce, particularly in the language domain.

Also, look at the work you do each day and consider "what will be automated" and "what could I do if I didn't need to do this?" The answers that you come up with will help you figure out how work will be distributed between you and bots, and what your job might look like when they arrive.

Finally, pretend that you don't your line manager no longer exists, and ask yourself honestly "what is it that they do that I will miss". If you have a long list congratulations you have a great manager and there are probably a few things on here that you will still need a human manager for (whatever their role will be called). If your list is really short then chances are they won't be around for long.

- **Do you think you need to do anything to prepare for a future in which bots will be coworkers?**
- **If you are a middle manager, do you have any plans to prepare for a future where your job will change substantially or disappear?**

- **On balance, do you think the future of work will be better or worse for employees?**
- **Are there any aspects of the future of work that trouble you?**
- **Will you advocate that change happen in a positive direction in which people are assisted with rather than dominated by technology?**

Consolidated list of questions for reflection

SECTION ONE - WORK HAS CHANGED FOREVER

The last gasp of industrial age management

- *How effective are the management practices in your workplace, and what actual evidence do you have to support your view?*
- *Are your management models imported from another era?*
- *What would happen if all your managers stopped coming to work?*

Working from home has sparked changes to work practices

- *Do your people require real-time connection with one another, and do they have it?*
- *Can you see parallels from other businesses types that have been operating in remote and distributed ways for years?*
- *Are there elements of your business that will never change because of remote work? What makes them so insulated from it?*
- *What is the optimal balance between remote, colocated or hybrid models. How would you know if it needed changing?*

The rise of AI and automation

- *Do you see applications AI in your personal life more so than in your work/professional life? Why is that?*
- *Is AI a fad? Do people treat it as such in your business?*
- *Do you have any views about how AI will change work in your organisation? What about your own work?*

SECTION TWO - MANAGERS ARE VULNERABLE TOO

Under threat

- *Do the managers in your organisation get the best out of your organisations investment in talent? Who is expected to do that?*
- *Do managers have too many other duties to effectively manage their people, and are they the right people to do this?*
- *How are middle managers coping with work since the COVID pandemic?*
- *What effort or investment goes into supporting management?*

- *Are managers leading change or resisting it, and what is being done about that?*

The mathematics of replacement

- *Have you considered the cost of management activities in your organisation, and whether they add value?*
- *Do you believe that management activities in your business are efficient or inefficient?*
- *What is the ratio of managers to workers? Is that appropriate for your organisation, and how do you know?*
- *Does your business have management bloat?*
- *Are decisions made with nimbleness, and if not is that evidence of bloat?*

Managers are struggling to do the 'human' things anyway

- *What do the worker in your business think of their management? Do they perceive it as highly functional, and worthy of trust?*
- *What kind of impact do you think the managers have on the workforce, and does this impact support your organisation's goals?*
- *Is your management exclusively focussed on the tasks and decisions of work? Do they get involved in improving how people do their work?*

Management will eat itself

- *What are the management tasks that can only be performed by a manager?*
- *How autonomously can your teams operate? How do you know?*
- *Are your teams getting the things that only human manager can provide from them?*

SECTION THREE - CONVERSATIONAL BOTS

Where conversations happen

- *How many software applications do your employees have to use to get their jobs done?*
- *Is your business adopting systems that work together cohesively, or are they still disconnected?*
- *Is the data generated by your staff as they work accessible from your current analysis platforms?*

- *Is anyone looking after the full digital experience for employees?*
- *Have you considered how conversational bots might offer an alternative user interface for users?*

Bots are finding jobs

- *Where are bots currently working in your office?*
- *Can you see natural applications for the use of bots in your workplace?*
- *Do you think you could augment your workforce with some form of automation, and if so do you have a transition plan?*

Under the hood: how bots work

- *Does the idea that bots will have such a broad range of capabilities concern you, and if so, why?*
- *Do you think bots are a fad, or is there something you should do to prepare for their growth and adoption?*

The bots are getting better

- *If bots continue to increase their capabilities, what kind of impact do you think that they will have on the workplace?*
- *Are there any things that people do that you think a bot will never be able to do?*
- *Would you like to see a bot try a creative task?*

But bots can't climb stairs

- *What is your experience at this point of interacting with bots?*
- *Have you been impressed or annoyed by them and what is it about them that has caused this?*
- *Do you feel that people are expecting too much of bots? Are they intimidated by them?*

Botty training

- *Are bots perceived as a predominantly technical capability that sits with IT ownership within your organisation?*
- *If individuals were free and able to create bots that enabled them to handle aspects of their job in different ways, do you think they would be open to doing so?*

SECTION FOUR - PEOPLE ANALYTICS

The emergence of 'People Analytics'

- *How well is your organisation's investment in people and how they use their time, talent and energy measured?*
- *Do you have any metrics that you can rely on to tell you this answer?*
- *What data about your workforce can't you live without?*
- *What do you wish you could know about your workforce that you can't get an answer to?*
- *Do larger organisational decisions genuinely factor the the impact on employees' work and the downstream costs that might be incurred?*

The tsunami of data in the modern workplace

- *Do you know your customers better than your people?*
- *Do you know what data is created at work, and what value it might be able to add?*
- *When was the last time you looked at this data or imagined how it might be used?*
- *Who has access to the rich data sets being created by your workers as they do their jobs?*

New forms of data

- *Have you optimised any processes using this new form of data?*
- *Are surveys or perceived as a problem by your workforce?*
- *Have you looked for different ways of accessing and interpreting 'employee voice'*

People Analytics in action

- *Are there any mistakes or errors that are continually repeated in your organisation? Why?*
- *Have you crystallised why certain practices seem to work or have a method of repeating success?*
- *Do you or your organisation cling to any sacred cows or dogma that you suspect might not be right?*
- *What data analysis initiatives do you have underway right now focused on discovering areas for business improvement.*

People Analytics inaction

- *Can you actually act on newly generated 'people' insight with your workforce or is that a 'no-go' zone?*
- *Do you expect decisions about how management should be practiced to have as much justification as decisions about other expenditures?*
- *Do you have the tools or technology and the kinds of skills that will allow for analysis of the critical drivers of value from your people?*
- *Does your organisation have access to the data it creates as it does work, or does that sit out of reach? Do you have data systems that relate to your employees that cannot be connected with other data for analysis?*
- *Have you tried integrating your various data sources that may shine a light on how well your people investment is working, or are you relying on a set of independent reports and out of date metrics?*
- *When confronted with a question about the workforce, do you sometimes hear that your HR system doesn't have that data, or support its analysis?*
- *Have your teams been seeking balanced investment across skills, analysis tools, and systems, or just in individual transactional HR systems?*
- *Is your workforce analytics capability funded to the same degree that other analysis functions are? If not, why not?*
- *Do you have a culture of testing and seeing what is possible?*
- *Has the case already been made that how people work is a critical success factor, and should therefore be analysed?*
- *Does all the investment in analytics get diverted to other parts of the business?*

An enterprise framework for People Analytics

- *Are you able to link people's behavior to business outcomes empirically, or does it happen purely on an intuitive level?*
- *Are you using data to adapt your management practices?*
- *Do you measure the impact of 'people' initiatives on business outcomes?*

SECTION FIVE - AGILE HUMAN RESOURCE MANAGEMENT

Water bears and viruses

- *Does your business understand its vulnerabilities and weaknesses?*
- *Are you evolving as an organisation, and if so, are you getting more 'fit' for what you expect the future will hold?*
- *Do you have any apparatus to discover whether your transformation is making you more fit or less fit?*
- *Is there anything that can be improved to help increase the velocity of adaptation?*

Every moment is an experiment

- *Does how well your organisation is working matter and if so does your business invest enough in understanding this?*
- *Is there anything that you are planning on rolling out soon that you could do a small, quick or inexpensive test to validate first?*
- *How often does the appeal to 'best-practice', or 'personal experience' win an argument in your organisation? How often do people appeal to data derived from your own organisations experience?*
- *Do new ideas for how your people work get suggested, and tested?*
- *Do managers in your organisation consider how effective they are in their work? Are they continuously improving? How do they know?*

Agile learning

- *Are people in your organisation suggesting that you need to be more agile?*
- *If so, are they using agile as an adjective for how you should operate, or as a noun, signify a framework they want you to invest in?*
- *How does you business record lessons learned, if at all?*
- *Is there a body of knowledge that you use, and if so, do you contribute to it?*
- *When considering change, does your organisation assess it from what it has already tried?*

Some worthwhile experiments

- *When was the last time you stopped a management practice because you had evidence it wasn't working?*

- *Does your organisation actively consider experiments that it can conduct to learn better methods?*
- *What about your workplace is the most surprising productivity enhancer that you have discovered?*

A growth hacking mindset

- *Which channels do your employees notice? Which ones do you use to influence them? Is it working?*
- *Are your leader's announcements seen as a distraction by the workforce?*
- *What can your leadership do differently to influence your people in a low-cost way?*
- *Given a choice, how do your workers prefer to connect?*

Employee-centred design

- *Does your business prioritise the employee experience when designing change*
- *Is employee experience part of your organisations' talent strategy*
- *Have methods for involving people in design and change programs been properly thought through, or is lip-service being paid instead?*

SECTION SIX - THE CONVERSATIONAL ENTERPRISE

Listen, Learn, Act

- *Does information about the workplace flow through your organisation*
- *Is your business relying on middle management to connect parts of your business?*

Displacement or augmentation

- *Are the executive in your organisation too busy to engage with the front-line workers to the degree they'd like to?*
- *What is it that your managers do at work? Are they coordinating people, developing them, busy with their own tasks, analysing performance, recommending change?*
- *What are the distinctive qualities that your workplace expects of its managers? Are they fulfilling them?*
- *What would happen if your managers stopped coming to work?*

The new roles needed

- *Does your business invest in coaching its people?*
- *If bots became more widespread in your workplace - who would look after them?*

The Org Chart of the Conversational Enterprise

- *Do your managers provide a consistent or inconsistent management experience? Does that matter?*
- *Is the control structure of the organisation working?*
- *Are managers an effective way of coordinating and cascading communications, or does the current approach introduce inconsistency?*
- *Is your structure too complicated? Can it cope simply with numeric growth?*
- *What, if anything, would be the barriers to supporting managers with coaching-Aides?*

SECTION SEVEN - THE CONVERSATIONAL ENTERPRISE GONE WRONG

Surveillance escalates

- *Are there kinds of analysis that you would feel personally uncomfortable if they were happening in your organisation?*
- *Were you in the 'let's take a peek' or in the 'must resist' camp when it came to gazing into the crystal ball and knowing everything?*
- *How does context change what you are willing to do with data?*
- *Do you believe that business leaders have explored these questions sufficiently enough to be making decisions for others about them?*

The 'Ideal Prison' and the workplace

- *Have you considered the trust implications of analysing the behavioral data of your employees?*
- *Is there any analysis that you believe would cross the line of acceptability / if so, what is it? What would be the implications of doing it?*

Turning employees into adversaries

- *What do you think your staff would expect your organisation to know about them?*

- *Do the managers in your organisation invest time in personally knowing all the workers, or is it just limited to specific friends?*
- *Is how you analyze workers' data congruent/incongruent with what you say about company values?*

Building and eroding trust

- *Does your organisation have a policy that defines how it uses internal data that is linked with employees? If so, do people understand it?*
- *If you have the power, would you let people see the data that might be used for decision making about them? If not, why not?*

Squashing diversity

- *Do you know what 'good' looks like in your organisation with respect to the behaviors of high performers?*
- *Does your organisation's you want to maintain a diversity of thought, style, approach, and background? Are there universal behaviours that you want everyone to share?*
- *Are there any management practices that currently stifle diversity?*

Bots in the wrong places

- *Would you expect people to mistreat a bot, and do you think the behavior could transfer back to groups of people?*
- *Do you think using a bot to handle common common queries make people feel devalued? Is it any different from publishing a page of information?*
- *What do you think the risks would be of making everyone talk to a bot?*

SECTION EIGHT - ALTERED RITUALS

Hiring better

- *Have you recently adjusted your hiring practices in response to the change in the nature of work, and access to talent?*
- *Do you have data to show which elements of your practices are working, and which need tweaking?*
- *Have you considered how prospective employees might evaluate your organisation in the context of global, work anywhere arrangements? Will you be competitive?*

Continuous onboarding

- *Does your organisation survive on the basis that people just know who they need to talk to to get stuff done?*
- *Is there a strong culture of documentation within your organisation?*
- *Can people find what they need, when they need it?*
- *Do your processes support the productive integration of new people in the environment?*
- *Is the ongoing development of staff something your businesses is already doing, and if so are you satisfied with the results of the investment in it?*
- *Are you able to measure understanding and adoption of practices and policies?*

Performance Management vs Managing Performance

- *Is it easy to know how effectively people are contributing to work outcomes?*
- *Does your business have any processes intended to help increase performance that are a waste of time? If so why are they retained?*
- *Who is evaluating performance (if anyone) and are they in a position to have a realistic basis for doing so. Who is best placed to provide input on performance?*
- *Are rewards and opportunities given on the basis of good performance, and if so is this just creating a situation of bias and widespread animosity?*

Nudges galore

- *When your organisation is trying to influence how people go about their work, is its first reaction to tell or to nudge?*
- *Have you implemented any nudges in your digital channels that may be experienced remotely?*
- *The impact of nudges is measurable - do you have the means of measuring them in place?*

Bots can have tough conversations too

- *Do your managers avoid having the hard conversations with staff?*
- *Can you see a role for bots assisting in conflict avoidance?*

Reducing organisational drag

- *Are people's days becoming increasing full of distractions from work, or are they now able to focus more on work?*
- *Do management activities increase or decrease the burden of work on individuals?*
- *What would be the impact of stopping status 'check-ins' in your workplace?*
- *Do people spend to long finding answers to questions and working out who can help them?*
- *Is communication efficient?*

Business planning

- *Is your organisation able to get useful input into organisational planning from your workforce? If not, what is stopping you?*
- *Would it be beneficial to have an automated process for listening to employees, and continuously synthesising this input for ongoing planning?*

Policy and Compliance Management

- *Do people understand your policies? Do they apply them?*
- *Is it reasonable to expect that your workers will be across all the details of relevant policies as they change?*
- *How many policies do you have?*
- *Is it easy for workers to discover whether any policies impact what they are planning on doing?*
- *Do you have the right policies, and do they support your actual business needs?*
- *Does your business continue to collect feedback and suggestions about your policies?*
- *Do you have the right policies? How do you know?*

Risk, Incidents, Health and Safety

- *Will algorithms ever be connected to employee health data? Is there a scenario in which employees would welcome this?*
- *Are there any real difference between monitoring on-the-job activities (such as location and movement), and other wearable data such as blood pressure and heart rate?*
- *Is it easy for workers to raise risks in a form that they are visible and handled?*

- *Does your business continue to collect feedback and suggestions about your risks and controls?*
- *Could a bot do a good job of managing the response to an incident? If so what kinds of incidents might they struggle with.*

Employer of choice

- *Is it important to your business that your workers regard your organisation highly?*
- *Has your organisation been using flexibility or the workplace environment are a strategy to attract and retain talent?*
- *Do you have a plan and the means for knowing why people join and stay in the future?*
- *Have you evaluated the impact of remote work on the employee experience? How do you know whether you are providing a compelling employee experience?*
- *To what lengths do you want to cater to the needs and desires of workers?*
- *Does your workplace appeal to any specific groups of workers with different needs, and if so why?*

I'm WFW tomorrow

- *If your office was permanently unavailable what would be the impact on your people?*
- *Are there things you would do to compensate for its absence? Are you already doing these things?*

Hiring teams rather than individuals

- *If you had the opportunity to do so, would you hire pre-existing teams, and what qualities would you look for in them?*
- *How do you think team-based incentives would change performance management practices?*
- *Do you know where you would go to source a team to help you?*

Who is actually doing the work?

- *Have any of your staff taken on work outside of their duties since working remotely?*
- *Do you believe that workers in your business are actually engaged in meaningful work, or are they filling time in return for pay?*

- *Are the managers at your work aware of the reality of what their people do?*
- *If you accidentally hired a productive bot instead of a person, what that bother you? Why?*
- *What would be the impact in the workplace if someone grew emotionally attached to a bot?*

Changing the culture

- *Is maintaining a certain culture a useful and realistic goal for a distributed workforce?*
- *Is your aspirational culture a good fit for your mission, and your people? How do you know if that is the case?*
- *What can be done to shift the culture in a positive direction, and how would anyone know if those attempts were working?*

Diversity and Inclusion

- *Should diversity and inclusion reporting include day-to-day operational activities, or just major employee lifecycle moments?*

Evolving the structure

- *What has motivated any restructure that you have been part of or initiated in the past?*
- *Did it get the results that were being sought?*
- *Do you know of any methods by which organisation structure can be designed that link it with desired business outcomes?*

SECTION NINE - PREPARING FOR THE FUTURE

- *Do you think you need to do anything to prepare for a future in which bots will be coworkers?*
- *If you are a middle manager, do you have any plans to prepare for a future where your job will change substantially or disappear?*
- *On balance, do you think the future of work will be better or worse for employees?*
- *Are there any aspects of the future of work that trouble you?*
- *Will you advocate that change happen in a positive direction in which people are assisted with rather than dominated by technology?*

Bibliography

"AlphaGo: The story so far." Deepmind, /research/case-studies/alphago-the-story-so-far.

Akshara. "How Lawyer Bots Are Revolutionizing the Legal Field." *The Black Box*, 23 Aug. 2020, https://medium.com/the-black-box/how-lawyer-bots-are-revolutionizing-the-legal-field-70b95044f9a8.

Arellano, Carla, et al. Using People Analytics to Drive Business Performance: A Case Study | McKinsey. 2017, https://www.mckinsey.com/business-functions/mckinsey-analytics/our-insights/using-people-analytics-to-drive-business-performance-a-case-study.

Arkcoll, Philip. Update: The Impact of COVID-19/WFH on Work Habits. https://www.linkedin.com/pulse/update-impact-covid-19wfh-work-habits-philip-arkcoll.

Arthur, Charles, and technology editor. "Yahoo Chief Bans Working from Home." The Guardian, 25 Feb. 2013. www.theguardian.com, https://www.theguardian.com/technology/2013/feb/25/yahoo-chief-bans-working-home.

Aufray, Jonathan. "Growth Hacking Definition: The Definitive One." Growth Hackers, 12 Sept. 2016, https://www.growth-hackers.net/growth-hacking-definition/.

Baig, et al. *Digital Adoption through COVID-19 and beyond | McKinsey*. https://www.mckinsey.com/business-functions/mckinsey-digital/our-insights/the-covid-19-recovery-will-be-digital-a-plan-for-the-first-90-days.

Bailie, Ian. "How Can HR Tech Help Users to Own Their Data?" MyHRfuture, 2019, https://www.myhrfuture.com/blog/2019/2/17/how-can-hr-tech-help-users-to-own-their-data.

Beauchene, Vinciane, and Molly Cunningham. "The End of Management as We Know It." *ANZ - EN*, 17 July 2020, https://www.bcg.com/en-au/publications/2020/end-management-as-we-know-it.

Beck, Randall, and James Harter. "Why Good Managers Are So Rare." Harvard Business Review, Mar. 2014. hbr.org, https://hbr.org/2014/03/why-good-managers-are-so-rare.

Beer, Michael, et al. "Why Leadership Training Fails—and What to Do About It." Harvard Business Review, Oct. 2016. hbr.org, https://hbr.org/2016/10/why-leadership-training-fails-and-what-to-do-about-it.

Belvedere, Matthew J. "CEOs Are at Their Wits' End – They Don't Know How to Get Their Employees Back in." *CNBC*, 1 Oct. 2021, https://www.cnbc.com/2021/10/01/ceos-are-at-their-wits-ends-dont-know-how-to-get-employees-back-in.html.

Ben-Ner, Avner, et al. "Treadmill Workstations: The Effects of Walking While Working on Physical Activity and Work Performance." PLoS ONE, edited by Robert L. Newton, vol. 9, no. 2, Feb. 2014, p. e88620. DOI.org (Crossref), doi:10.1371/journal.pone.0088620.

Berasategi, Ane. "Sarcasm Detection with NLP." Medium, 25 Jan. 2020, https://towardsdatascience.com/sarcasm-detection-with-nlp-cbff1723f69a.

Bersin, Josh. Employee Engagement 3.0 - From Feedback to Action. 14 Apr. 2019, https://joshbersin.com/2019/04/employee-engagement-3-0-from-feedback-to-action/.

---. "People Analytics and AI in the Workplace: Four Dimensions of Trust." JOSH BERSIN, 4 May 2019, https://joshbersin.com/2019/05/the-ethics-of-ai-and-people-analytics-four-dimensions-of-trust/.

---. People Analytics: Here With A Vengeance. 19 Dec. 2017, https://joshbersin.-com/2017/12/people-analytics-here-with-a-vengeance/.

---. "The Geeks Arrive In HR: People Analytics Is Here." Forbes, Feb. 2015, https://www.forbes.com/sites/joshbersin/2015/02/01/geeks-arrive-in-hr-people-analytics-is-here/

—-. "Is The Employee Survey Dead? Nope. It's Becoming Smarter By The Minute." JOSH BERSIN, 2 Aug. 2020, https://joshbersin.com/2020/08/is-the-employee-survey-dead-nope-its-becoming-smarter-by-the-minute/.

"Better Language Models and Their Implications." OpenAI, 14 Feb. 2019, https://openai.com/blog/better-language-models/.

Benzinga. "Amazon's Warehouse Productivity Measure Sparks New California Law." *CFO*, 23 Sept. 2021, https://www.cfo.com/workplace-issues/2021/09/amazons-warehouse-productivity-measure-sparks-new-california-law/.

Bock, Laszlo. Work Rules! Insights from inside Google That Will Transform How You Live and Lead. First edition, Twelve, 2015.

Brooks, Frederick P. The Mythical Man-Month: Essays on Software Engineering. Addison-Wesley Pub. Co, 1975.

Brown, Gavin, and Richard Whittle. Algorithms, Blockchain et Cryptocurrency: Implications for the Future of the Workplace. 2020. Open WorldCat, https://doi.org/10.1108/9781838674953.

Cain, Áine. "Jeff Bezos's Productivity Tip? The '2 Pizza Rule.'" Inc.Com, 7 June 2017, https://www.inc.com/business-insider/jeff-bezos-productivity-tip-two-pizza-rule.html.

Campbell, Charlie. "The Entire System Is Designed to Suppress Us': What the Chinese Surveillance State Means for the Rest of the World." Time, Nov. 2019, https://time.com/5735411/china-surveillance-privacy-issues/.

Capelli, and Eldor. "Where Measuring Engagement Goes Wrong." Harvard Business Review, May 2019. hbr.org, https://hbr.org/2019/05/where-measuring-engagement-goes-wrong.

Cappelli, Peter. "It's Time to Get Rid of Employee Surveys." Wall Street Journal, 1 Aug. 2020. www.wsj.com, https://www.wsj.com/articles/its-time-to-get-rid-of-employee-surveys-11596246683.

Capgemini. "Cultural Issues Are the Number One Obstacle to Digital Transformation as Digital Cultural Divide Widens by 7% Points Compared to Six Years Ago." Capgemini Australia, 8 June 2017, https://www.capgemini.com/au-en/news/cultural-issues-are-the-number-one-obstacle-to-digital-transformation-as-digital-cultural-divide-widens-by-7-points-compared-to-six-years-ago/.

Chamorro-Premuzic, Tomas, et al. “Building Ethical AI for Talent Management.” Harvard Business Review, Nov. 2019. hbr.org, https://hbr.org/2019/11/building-ethical-ai-for-talent-management.

Chemi, Jessica Golden, Eric. “Worker Monitoring Tools See Surging Growth as Companies Adjust to Stay-at-Home Orders.” CNBC, 13 May 2020, https://www.cnbc.com/2020/05/13/employee-monitoring-tools-see-uptick-as-more-people-work-from-home.html.

Cheremond, RJ. *6 Ways the Workplace Will Change in the Next 10 Years.* //www.gartner.com/smarterwithgartner/6-ways-the-workplace-will-change-in-the-next-10-years/.

Chui, Michael, et al. The Social Economy: Unlocking Value and Productivity through Social Technologies. https://www.mckinsey.com/industries/technology-media-and-telecommunications/our-insights/the-social-economy#.

CIPD. Rise in Stress at Work Linked to Poor Management, New Research Finds | CIPD Press Release. https://www.cipd.co.uk/about/media/press/rise-stress-poor-management.

Coolen, Patrick, and Luuk Sombezki. “The 10 Golden Rules of Continuous Employee Listening.” The 10 Golden Rules of Continuous Employee Listening, May 2019, https://www.linkedin.com/pulse/10-golden-rules-continuous-employee-listening-patrick-coolen/.

Coolen, Patrick, and Frank van den Brink (May 2018) . HR Is Hitting a Second Wall. May 2018, https://www.linkedin.com/pulse/hr-hitting-second-wall-patrick-coolen.

Cooley, Mike. “Human-Centred Systems.” Designing Human-Centred Technology, edited by Howard Rosenbrock, Springer London, 1989, pp. 133-43. DOI.org (Crossref), doi:10.1007/978-1-4471-1717-9_10.

Curtin, Melanie. “The Resume Is Dead. Here’s What Innovative Companies (Including Tesla) Are Using to Hire Instead.” Inc.Com, 27 June 2018, https://www.inc.com/melanie-curtin/resumes-dont-help-you-hire-innovative-people-but-this-does-hint-teslas-doing-it.html.

Damien Dellala. People Analytics Foundation. https://www.slideshare.net/ddellala/people-analytics-foundation.

Das Swain, Vedant, et al. “A Multisensor Person-Centered Approach to Understand the Role of Daily Activities in Job Performance with Organisational Personas.” Proceedings of the ACM on Interactive, Mobile, Wearable and Ubiquitous Technologies, vol. 3, no. 4, Dec. 2019, pp. 1-27. DOI.org (Crossref), doi:10.1145/3369828.

Davenport, Thomas H., and Jeanne G. Harris. Competing on Analytics: The New Science of Winning. Harvard Business School Press, 2007.

David Green. “MyHRfuture.” Blog, https://www.myhrfuture.com/blog.

de Jesus, Ana Carolina C., et al. “Exploiting Linkedin to Predict Employee Resignation Likelihood.” Proceedings of the 33rd Annual ACM Symposium on Applied Computing - SAC ’18, ACM Press, 2018, pp. 1764-71. DOI.org (Crossref), doi:10.1145/3167132.3167320.

Deloitte. The Role of Culture in Digital Transformation. https://deloitte.wsj.com/cio/2019/07/18/the-role-of-culture-in-digital-transformation/.

Deutsch, and Sterling. "Uber Drivers Are Employees, Not Contractors, Says Dutch Court." *Reuters*, 13 Sept. 2021, https://www.reuters.com/world/europe/dutch-court-rules-uber-drivers-are-employees-not-contractors-newspaper-2021-09-13/.

Dhar, Julia, et al. "The Persuasive Power of the Digital Nudge." ANZ - EN, 2017, https://www.bcg.com/en-au/publications/2017/people-organisation-operations-persuasive-power-digital-nudge.

Diallo. "The Machine Fired Me." *Ibrahim Diallo Blog*, https://idiallo.com/blog/when-a-machine-fired-me.

Dick, Philip K. Blade Runner (Do Androids Dream of Electric Sheep). Random House, 1968.

Doerr, John E. Measure What Matters: How Google, Bono, and the Gates Foundation Rock the World with OKRs. Portfolio/Penguin, 2018.

Drucker, Peter F. The Practice of Management. HarperCollins, 2009.

Elgammal, Ahmed. "AI Is Blurring the Definition of Artist." American Scientist, 6 Dec. 2018, https://www.americanscientist.org/article/ai-is-blurring-the-definition-of-artist.

Fenner, Robert (2018) "Alibaba's AI Outguns Humans in Reading Test" Bloomberg, https://www.bloomberg.com/news/articles/2018-01-15/alibaba-s-ai-outgunned-humans-in-key-stanford-reading-test.

Fickman, Laurie. *Using Your Smartwatch to Reduce Stress*. https://www.uh.edu/news-events/stories/august-2021/08242021-faghih-smartwatch-detects-stress.

Fix, Blair. "The Growth of Managers: An Update." Economics from the Top Down, 23 May 2019, https://economicsfromthetopdown.com/2019/05/23/the-growth-of-managers-an-update/.

Ford, Neal, et al. Building Evolutionary Architectures: Support Constant Change. O'Reilly Media, 2017.

"Forgetting Curve." Wikipedia, 15 July 2020. Wikipedia, https://en.wikipedia.org/w/index.php?title=Forgetting_curve&oldid=967823302.

Frey, Carl Benedikt, and Michael Osborne. "The Future of Employment: How Susceptible Are Jobs to Computerisation?" *Oxford Martin School*, https://www.oxfordmartin.ox.ac.uk/publications/the-future-of-employment/.

Fried, Jason, and David Heinemeier Hansson. Remote: Office Not Required. 1st edition, Crown Business, 2013.

---. Rework. 1st ed, Crown Business, 2010.

Gallup, Inc. "Managers Account for 70% of Variance in Employee Engagement." Gallup.Com, 21 Apr. 2015, https://news.gallup.com/businessjournal/182792/managers-account-variance-employee-engagement.aspx.

Galston, William A. "Is Seeing Still Believing? The Deepfake Challenge to Truth in Politics." Brookings, 8 Jan. 2020, https://www.brookings.edu/research/is-seeing-still-believing-the-deepfake-challenge-to-truth-in-politics/.

"General Data Protection Regulation (GDPR) - Official Legal Text." General Data Protection Regulation (GDPR), https://gdpr-info.eu/.

Gilbert, PRO Compliance-Samantha. Is a Federal US Data Protection Regime Closer than We Thought? | Lexology. https://www.lexology.com/library/detail.aspx?g=21821e38-0a05-4ca5-8d69-3ba2a34ab6f1.

GitLab. "GitLab's Guide to All-Remote." GitLab, https://about.gitlab.com/company/culture/all-remote/guide/.

Goasduff, Laurence. *Robobosses Enhance Management Capabilities*. //www.gartner.com/smarterwithgartner/robobosses-enhance-management-capabilities/.

Goleman, Daniel. Emotional Intelligence. 10th anniversary trade pbk. ed, Bantam Books, 2005.

---. Primal Leadership: Unleashing the Power of Emotional Intelligence. Tenth anniversary edition, Harvard Business Review Press, 2013.

GPT-3. "A Robot Wrote This Entire Article. Are You Scared yet, Human? | GPT-3." *The Guardian*, 8 Sept. 2020, http://www.theguardian.com/commentisfree/2020/sep/08/robot-wrote-this-article-gpt-3.

Grant, Adam. "The Surprising Value of Obvious Insights." MIT Sloan Management Review, https://sloanreview.mit.edu/article/the-surprising-value-of-obvious-insights/.

Green, David. COVID-19: The Role of HR & People Analytics in the Crisis. 3 May 2020, https://www.linkedin.com/pulse/covid-19-role-hr-people-analytics-crisis-david-green.

---. "People Analytics in 2020: Growth, Leading Practices, Case Studies and Ethics." MyHRfuture, Jan. 2020, https://www.myhrfuture.com/blog/2020/1/27/people-analytics-in-2020-growth-leading-practices-case-studies-and-ethics.

---. Putting the 'H' Back into HR - 10 Predictions for HR in 2020. Dec. 2019, https://www.linkedin.com/pulse/putting-h-back-hr-10-predictions-2020-david-green.

Greenwood, Kelly, and Natasha Krol. "8 Ways Managers Can Support Employees' Mental Health." Harvard Business Review, Aug. 2020. hbr.org, https://hbr.org/2020/08/8-ways-managers-can-support-employees-mental-health.

Griffin, De'Onn, and Mark Coleman. "How We Will Work in 2028." *Gartner*, https://www.gartner.com/en/documents/3861479/how-we-will-work-in-20280.

Groenfeldt, Tom. "Financial Regulations Will Surpass 300 Million Pages by 2020 Says JWG." Fintech - News and Analysis, 20 Apr. 2016, https://techandfinance.com/2016/04/20/financial-regulations-will-surpass-300-million-pages-by-2020-says-jwg/.

"Growth Hacking." Wikipedia, 20 July 2020. Wikipedia, https://en.wikipedia.org/w/index.php?title=Growth_hacking&oldid=968677602.

Guenole, Nigel, et al. The Power of People: Learn How Successful Organisations Use Workforce Analytics to Improve Business Performance. Pearson Education, 2017.

Gupta, Aparna, and Abena Owusu. "Identifying the Risk Culture of Banks Using Machine Learning." SSRN Electronic Journal, 2019. DOI.org (Crossref), doi:10.2139/ssrn.3441861.

Haak, Tom. "Nearly 100 Leadership Models." HR Trend Institute, 24 Feb. 2020, https://hrtrendinstitute.com/2020/02/24/nearly-100-leadership-models/.

Hamel, Gary. "First, Let's Fire All the Managers." Harvard Business Review, no. December 2011, Dec. 2011. hbr.org, https://hbr.org/2011/12/first-lets-fire-all-the-managers.

Hamori, Monika. Why Job-Hopping Is on the Rise. 2010, https://www.hcamag.-com/au/news/general/why-job-hopping-is-on-the-rise/140364.

Hao, Karen. "China Has Started a Grand Experiment in AI Education. It Could Reshape How the World Learns." MIT Technology Review, Aug. 2019, https://www.technologyreview.com/2019/08/02/131198/china-squirrel-has-started-a-grand-experiment-in-ai-education-it-could-reshape-how-the/.

---. "OpenAI's Fiction-Spewing AI Is Learning to Generate Images." MIT Technology Review, July 2020, https://www.technologyreview.com/2020/07/16/1005284/openai-ai-gpt-2-generates-images/.

---. "This Is How AI Bias Really Happens—and Why It's so Hard to Fix." MIT Technology Review, Feb. 2019, https://www.technologyreview.com/2019/02/04/137602/this-is-how-ai-bias-really-happensand-why-its-so-hard-to-fix/.

Harbinja, Edina, et al. "Chatbots That Resurrect the Dead: Legal Experts Weigh in on 'Disturbing' Technology." *The Conversation*, http://theconversation.-com/chatbots-that-resurrect-the-dead-legal-experts-weigh-in-on-disturbing-technology-155436.

Hastings, Reed. Culture. https://www.slideshare.net/reed2001/culture-1798664.

Heskett, James L., et al. "Putting the Service-Profit Chain to Work." Harvard Business Review, no. July-August 2008, July 2008. hbr.org, https://hbr.org/2008/07/putting-the-service-profit-chain-to-work.

Hinds, Pamela, and Sara Kiesler, editors. Distributed Work. MIT Press, 2002.

Hirsen, Adam. "Council Post: Why Your Business's 2019 Resolution Should Be Quitting Interruption Marketing." Forbes, https://www.forbes.com/sites/forbeschicagocouncil/2019/01/23/why-your-businesss-2019-resolution-should-be-quitting-interruption-marketing/.

How Airbnb Became the Largest Hotel Chain. https://www.eventige.com/blog/airbnb-largest-hotel-chain.

"How I Work." Amantha Imber, https://www.amanthaimber.com/podcast/.

Hutson, Matthew. "This 'Mind-Reading' Algorithm Can Decode the Pictures in Your Head." Science | AAAS, 10 Jan. 2018, https://www.sciencemag.org/news/2018/01/mind-reading-algorithm-can-decode-pictures-your-head.

Ibarra, Herminia, and Anne Scoular. "The Leader as Coach." Harvard Business Review, no. November-December 2019, Nov. 2019. hbr.org, https://hbr.org/2019/11/the-leader-as-coach.

Ingham, Jon. The Social Organisation: Developing Employee Connections and Relationships for Improved Business Performance. 1st Edition, Kogan Page Ltd, 2017.

"Introduction to Viruses." Wikipedia, 2 July 2020. Wikipedia, https://en.wikipedia.org/w/index.php?title=Introduction_to_viruses&oldid=965637999.

Isson, Jean Paul, and Jesse Harriott. People Analytics in the Era of Big Data: Changing the Way You Attract, Acquire, Develop, and Retain Talent. 1st ed., Wiley, 2016.

Jahn, Jens, et al. "Unlocking Change Management with People Analytics." ANZ - EN, 2019, https://www.bcg.com/en-au/publications/2019/change-management-with-people-analytics.

Jeffries, Ronald. *Developers Should Abandon Agile*. https://ronjeffries.com/articles/018-01ff/abandon-1/.

Jesuthasan, Ravin, and Suzanne McAndrew. "The Future Chief People Officer: Imagine. Invent. Ignite." Willis Towers Watson, 2020, https://www.willistowerswatson.com/en-US/Insights/2020/01/the-future-chief-people-officer-imagine-invent-ignite.

Judd, Scott, et al. "Employee Surveys Are Still One of the Best Ways to Measure Engagement." Harvard Business Review, Mar. 2018. hbr.org, https://hbr.org/2018/03/employee-surveys-are-still-one-of-the-best-ways-to-measure-engagement.

Kahn, William A. "Psychological Conditions of Personal Engagement and Disengagement at Work." Academy of Management Journal, vol. 33, no. 4, Dec. 1990, pp. 692–724. DOI.org (Crossref), doi:10.5465/256287.

Kejzlar, Tomas. "Cargo Cult Agile." Medium, 2 Sept. 2017, https://skepticalagile.com/cargo-cult-agile-23340e0647ae.

Khan, Nadeem, and Dave Millner. Introduction to People Analytics: A Practical Guide to Data-Driven HR. Kogan Page, 2020.

Kniberg, Henrik. "Spotify Engineering Culture (Part 1)." Spotify Engineering, 27 Mar. 2014, https://engineering.atspotify.com/2014/03/27/spotify-engineering-culture-part-1/.

Kollewe, Julia. "Google and Facebook Bring in One-Fifth of Global Ad Revenue." The Guardian, 1 May 2017. www.theguardian.com, https://www.theguardian.com/media/2017/may/02/google-and-facebook-bring-in-one-fifth-of-global-ad-revenue.

Korn Ferry Institute. Driving Team Effectiveness. 2016, https://www.kornferry.com/content/dam/kornferry/docs/article-migration/Korn-Ferry-Institute_Driving_team_effectiveness_2017_01.pdf.

Krotoski, Aleks. "BBC World Service - The Documentary, The Digital Human: Subservience." BBC, https://www.bbc.co.uk/programmes/w3ct1d0d.

Kruse, Kevin. "Top 5 Leadership Development Trends For 2020." *Forbes*, https://www.forbes.com/sites/kevinkruse/2020/01/15/top-5-leadership-development-trends-for-2020/.

Lencioni, Patrick. The Five Dysfunctions of a Team: A Leadership Fable. 1st ed, Jossey-Bass, 2002.

---. The Ideal Team Player: How to Recognize and Cultivate the Three Essential Virtues: A Leadership Fable. John Wiley and Sons, Inc, 2016.

—. The Motive: Why So Many Leaders Abdicate Their Most Important Responsibilities. 1st edition, Jossey-Bass, 2020.

Leonardi, Maria. "Water Cooler Talk – Innovation Silver Bullet or Facade?" *Management Matters*, 3 Feb. 2021, https://medium.com/management-matters/water-cooler-talk-innovation-silver-bullet-or-facade-7f1fad84de20.

Leusch, Yannik. Organisational Network Analysis - A Case Study. Sept. 2019, https://www.linkedin.com/pulse/organisational-network-analysis-case-study-yannik-leusch.

Lupshor, Stela. How People Analytics Can Drive New Business Insights. SAP Concur, 2019, https://www.concur.com/newsroom/article/how-people-analytics-can-drive-new-business-insights.

Mankins, Michael C., and Eric Garton. Time, Talent, Energy: Overcome Organisational Drag and Unleash Your Team's Productive Power. Harvard Business Review Press, 2017.

Manfredi, Lucas. "Microsoft CEO Warns against Permanent Work from Home." FOXBusiness, 17 May 2020, https://www.foxbusiness.com/technology/microsoft-ceo-permanent-work-from-home-warning.

Martin, Charlie. "Cargo Cult Methodology: How Agile Can Go Terribly, Terribly Wrong." CIO, 7 Aug. 2008, https://www.cio.com/article/2434415/cargo-cult-methodology--how-agile-can-go-terribly--terribly-wrong.html.

Martin, Fiona, and Jonathon Hutchinson. "Deep Data: Analyzing Power and Influence in Social Media Networks." Second International Handbook of Internet Research, edited by Jeremy Hunsinger et al., Springer Netherlands, 2020, pp. 857–77. DOI.org (Crossref), doi:10.1007/978-94-024-1555-1_19.

Maslow, A. H. "A Theory of Human Motivation." Psychological Review, vol. 50, no. 4, July 1943, pp. 370–96. DOI.org (Crossref), doi:10.1037/h0054346.

Mattingly, Stephen M., et al. "The Tesserae Project: Large-Scale, Longitudinal, In Situ, Multimodal Sensing of Information Workers." Extended Abstracts of the 2019 CHI Conference on Human Factors in Computing Systems, ACM, 2019, pp. 1–8. DOI.org (Crossref), doi:10.1145/3290607.3299041.

McChrystal, Stanley A., et al. Team of Teams: New Rules of Engagement for a Complex World. Portfolio/Penguin, 2015.

McDermid, Riley. Airbnb's Number of Listings Surpasses Rooms Held by Top 5 Hotel Brands Combined, Aug. 2017, https://www.bizjournals.com/sanfrancisco/news/2017/08/11/airbnb-surpasses-ihg-wyn-hilton-marriott-listings.html.

Mercer. 2020 Global Talent Trends Report | The Future of Work | Mercer. https://www.mercer.com/our-thinking/career/global-talent-hr-trends.html.

Mintzberg, Henry. The Nature of Managerial Work. Harper & Row, 1973.

Mirjafari, Shayan, et al. "Differentiating Higher and Lower Job Performers in the Workplace Using Mobile Sensing." Proceedings of the ACM on Interactive, Mobile, Wearable and Ubiquitous Technologies, vol. 3, no. 2, June 2019, pp. 1–24. DOI.org (Crossref), doi:10.1145/3328908.

Morieux, Yves. As Work Gets More Complex, 6 Rules to Simplify. 2017. www.ted.com, https://www.ted.com/talks/yves_morieux_as_work_gets_more_complex_6_rules_to_simplify.

---. How Too Many Rules at Work Keep You from Getting Things Done". 2017, https://www.ted.com/talks/yves_morieux_how_too_many_rules_at_work_keep_you_from_getting_things_done?language=my.

Murphy, Kevin R. "Performance Evaluation Will Not Die, but It Should." Human Resource Management Journal, vol. 30, no. 1, Jan. 2020, pp. 13-31. DOI.org (Crossref), doi:10.1111/1748-8583.12259.

myHRfuture. "The Best HR & People Analytics Articles of June 2020." MyHRfuture, https://www.myhrfuture.com/blog/2020/7/2/the-best-hr-amp-people-analytics-articles-of-june-2020.

Nichols, Greg. "Artificial Empathy: Call Center Employees Are Using Voice Analytics to Predict How You Feel." ZDNet, 2018, https://www.zdnet.com/article/artificial-empathy-call-center-employees-are-using-voice-analytics-to-predict-how-you-feel/.

Nikolic, Isabella. "Barclays Staff Slam Bosses for Using 'Big Brother' Software." Mail Online, 20 Feb. 2020, https://www.dailymail.co.uk/news/article-8025003/Barclays-staff-slam-bosses-using-Big-Brother-software-monitors-trips-toilet.html.

"OGSM." Wikipedia, 22 Apr. 2020. Wikipedia, https://en.wikipedia.org/w/index.php?title=OGSM&oldid=952453789.

"Open Letter on Artificial Intelligence." Wikipedia, 16 Feb. 2020. Wikipedia, https://en.wikipedia.org/w/index.php?title=Open_Letter_on_Artificial_Intelligence&oldid=941087355.

Parker, Sharon K., et al. "Remote Managers Are Having Trust Issues." Harvard Business Review, July 2020. hbr.org, https://hbr.org/2020/07/remote-managers-are-having-trust-issues.

Paycor. Labor Costs Are the Biggest Cost of Business. https://www.paycor.com/resource-center/a-closer-look-at-labor-costs. Accessed 21 July 2020.

Pentland, Alex "Sandy." "The New Science of Building Great Teams." Harvard Business Review, no. April 2012, Apr. 2012. hbr.org, https://hbr.org/2012/04/the-new-science-of-building-great-teams.

Percy, Sally. "The Five Big Advantages Of Working For A Robo-Boss." *Forbes*, https://www.forbes.com/sites/sallypercy/2019/10/21/the-five-big-advantages-of-working-for-a-robo-boss/. Accessed 2 Jan. 2021.

Perlow, Leslie A., and Jessica L. Porter. "Making Time Off Predictable–and Required." Harvard Business Review, no. October 2009, Oct. 2009. hbr.org, https://hbr.org/2009/10/making-time-off-predictable-and-required.

Porter, Leslie A. Perlow and Jessica L. "Making Time Off Predictable–and Required." Harvard Business Review, no. October 2009, Oct. 2009. hbr.org, https://hbr.org/2009/10/making-time-off-predictable-and-required.

Prachi, Juneja. The Role of Senior Managers as Barriers to Change. https://www.managementstudyguide.com/role-of-senior-managers-as-barriers-to-change.htm.

Press, Gil. "Is AI Going To Be A Jobs Killer? New Reports About The Future Of Work." Forbes, https://www.forbes.com/sites/gilpress/2019/07/15/is-ai-going-to-be-a-jobs-killer-new-reports-about-the-future-of-work/.

PYMNTS. "Reimagining The Bank As A Technology Company." PYMNTS.Com, 10 Apr. 2020, https://www.pymnts.com/news/digital-banking/2020/reimagining-bank-as-technology-company/.

Rasmussen, Thomas, and Dave Ulrich. "Learning from Practice: How HR Analytics Avoids Being a Management Fad." Organisational Dynamics, vol. 44, no. 3, July 2015, pp. 236–42. DOI.org (Crossref), doi:10.1016/j.orgdyn.2015.05.008.

"Reduce Organisational Drag." Harvard Business Review, Mar. 2017. hbr.org, https://hbr.org/podcast/2017/03/globalization-myth-and-reality-2.

Reichheld, Frederick F. "The One Number You Need to Grow." Harvard Business Review, no. December 2003, Dec. 2003. hbr.org, https://hbr.org/2003/12/the-one-number-you-need-to-grow.

Ries, Eric. The Lean Startup: How Constant Innovation Creates "Rise of Artificial Intelligence Means Architects Are 'Doomed' Says Sebastian Errazuriz." Dezeen, 22 Oct. 2019, https://www.dezeen.com/2019/10/22/artificial-intelligence-ai-architects-jobs-sebastian-errazuriz/.

Radically Successful Businesses. Portfolio Penguin, 2011.

Richardson, Tim. *IKEA Cites "health and Safety" for Hidden Cameras in Toilets.* https://www.theregister.com/2021/10/01/ikea_spycam_scandal/.

Rocca, Joseph. "Understanding Generative Adversarial Networks (GANs)." Medium, 19 July 2020, https://towardsdatascience.com/understanding-generative-adversarial-networks-gans-cd6e4651a29.

Röcker, Carsten. Informal Communication and Awareness in Virtual Teams: Why We Need Smart Technologies to Support Distributed Teamwork. 2012, https://interruptions.net/literature/Rocker-CISME12.pdf.

Rogers, Bruce. "Why 84% Of Companies Fail At Digital Transformation." Forbes, Jan. 2016, https://www.forbes.com/sites/brucerogers/2016/01/07/why-84-of-companies-fail-at-digital-transformation/.

Rozovsky, Julia. Re:Work - The Five Keys to a Successful Google Team. 2015, https://rework.withgoogle.com/blog/five-keys-to-a-successful-google-team/.

SAFETY DEVICE SAVING SPINES. Australian Government - Department of Industry, Innovation and Science, https://www.industry.gov.au/sites/default/files/growth-centres-safety-device-saving-spines.pdf.

Saha, Koustuv, et al. "LibRA: On LinkedIn Based Role Ambiguity and Its Relationship with Wellbeing and Job Performance." Proceedings of the ACM on Human-Computer Interaction, vol. 3, no. CSCW, Nov. 2019, pp. 1-30. DOI.org (Crossref), doi:10.1145/3359239.

Schiffer, Zoe. "Apple's Fortress of Secrecy Is Crumbling from the inside." *The Verge*, 30 Sept. 2021, https://www.theverge.com/22700898/apple-company-culture-change-secrecy-employee-unrest.

Schmidt, Adrien. "The Rise Of Conversational Interfaces And Their Impact On Business." *Forbes*, https://www.forbes.com/sites/theyec/2019/01/25/the-rise-of-conversational-interfaces-and-their-impact-on-business/.

Schotkamp, Tom, and Martin Danoesastro. "HR's Pioneering Role in Agile at ING." ANZ - EN, 2017, https://www.bcg.com/en-au/publications/2018/human-resources-pioneering-role-agile-ing.

Schwantes, Marcel. "Survey: 91 Percent of 1,000 Employees Say Their Bosses Lack This 1 Critical Skill." *Inc.Com*, 10 Aug. 2017, https://www.inc.com/marcel-schwantes/survey-91-percent-of-1000-employees-say-their-boss.html.

Seabrook, John. "Has the Pandemic Transformed the Office Forever?" *The New Yorker*, https://www.newyorker.com/magazine/2021/02/01/has-the-pandemic-transformed-the-office-forever.

Seppälä, Emma, and Kim Cameron. "Proof That Positive Work Cultures Are More Productive." *Harvard Business Review*, Dec. 2015. *hbr.org*, https://hbr.org/2015/12/proof-that-positive-work-cultures-are-more-productive.

Senz, Kristen. "How Much Will Remote Work Continue After the Pandemic?" *HBS Working Knowledge*, 24 Aug. 2020, http://hbswk.hbs.edu/item/how-much-will-remote-work-continue-after-the-pandemic.

Shacklett , Mary. "How AI and Voice Analytics Can Improve the Call Center Experience." TechRepublic, 11 Sept. 2019, https://www.techrepublic.com/article/how-ai-and-voice-analytics-can-improve-the-call-center-experience/.

Sheedy, Elizabeth, and Barbara Griffin. "Risk Governance, Structures, Culture, and Behavior: A View from the Inside." Corporate Governance: An International Review, vol. 26, no. 1, Jan. 2018, pp. 4–22. DOI.org (Crossref), doi:10.1111/corg.12200.

Shook, Ellyn, et al. Decoding Organisational DNA | Accenture. 2019, https://www.accenture.com/us-en/insights/future-workforce/workforce-data-organisational-dna.

Skillicorn, David B. "Adversarial Knowledge Discovery." IEEE Intelligent Systems, vol. 24, no. 6, Nov. 2009, pp. 54-61. DOI.org (Crossref), doi:10.1109/MIS.2009.108.

Soter Analytics. SOTER ANALYTICS HELPS THE TRAVIS PERKINS GROUPREDUCE MANUAL HANDLING INJURIES BY 55% . Feb. 2020, https://soteranalytics.com/wp-content/uploads/2020/06/soter_analytics_-_travis_perkins_-case_study_2019-2-1-1.pdf.

Spolsky, Joel. "The Joel Test: 12 Steps to Better Code." Joel on Software, 9 Aug. 2000, https://www.joelonsoftware.com/2000/08/09/the-joel-test-12-steps-to-better-code/.

Spolsky, Joel. "Smart and Gets Things Done." Joel on Software, 5 June 2007, https://www.joelonsoftware.com/2007/06/05/smart-and-gets-things-done/.

Storey, John, et al. Strategic Human Resource Management: A Research Overview. First Edition, Routledge, 2019.

Sutton, Robert I. "Bosses, Get Out of Your Employees' Way." *Wall Street Journal*, 28 Oct. 2019. *www.wsj.com*, https://www.wsj.com/articles/bosses-get-out-of-your-employees-way-11572228361.

Sullivan, Mark. "The Making of Mojo, AR Contact Lenses That Give Your Eyes Superpowers." Fast Company, 16 Jan. 2020, https://www.fastcompany.com/90441928/the-making-of-mojo-ar-contact-lenses-that-give-your-eyes-superpowers.

Thaler, Richard H., and Cass R. Sunstein. Nudge: Improving Decisions about Health, Wealth, and Happiness. Rev. and Expanded ed, Penguin Books, 2009.

"The Coronavirus Crisis Thrusts Corporate HR Chiefs into the Spotlight." The Economist. The Economist, https://www.economist.com/business/2020/03/24/the-coronavirus-crisis-thrusts-corporate-hr-chiefs-into-the-spotlight.

"The State of the Octoverse." *The State of the Octoverse*, https://octoverse.github.com/.

Tiersky, Howard. "5 Top Challenges to Digital Transformation in the Enterprise." CIO, 13 Mar. 2017, https://www.cio.com/article/3179607/5-top-challenges-to-digital-transformation-in-the-enterprise.html.

Twist. "Remote Work Guides." Twist, https://twist.com/remote-work-guides.

Ulrich, Dave. Recognizing and Overcoming Unconscious Biases in HR. 2019, https://www.linkedin.com/pulse/recognizing-overcoming-unconscious-biases-hr-dave-ulrich.

Ulrich, David, editor. HR from the Outside in: The next Era of Human Resources Transformation. McGraw-Hill, 2012.

Vincent, James. "OpenAI Has Published the Text-Generating AI It Said Was Too Dangerous to Share." The Verge, 7 Nov. 2019, https://www.theverge.com/2019/11/7/20953040/openai-text-generation-ai-gpt-2-full-model-release-1-5b-parameters.

Waber, Ben. People Analytics: How Social Sensing Technology Will Transform Business and What It Tells Us about the Future of Work. FT Press, 2013.

West, Stephen. Episode #121 ... Michel Foucault Pt. 1 - Discipline and Punish - YouTube. https://www.youtube.com/watch?v=HG_Q9WnOuxU.

Withers, Rachel. "I Don't Date Men Who Yell at Alexa." Slate Magazine, 30 Apr. 2018, https://slate.com/technology/2018/04/i-judge-men-based-on-how-they-talk-to-the-amazon-echos-alexa.html.

Wilkie, Dana. "Employee Engagement Surveys: Why Do Workers Distrust Them?" SHRM, 5 Jan. 2018, https://www.shrm.org/resourcesandtools/hr-topics/employee-relations/pages/employee-engagement-surveys.aspx.

WEF. "The Business Case For Diversity Is Now Overwhelming. Here's Why." *World Economic Forum*, https://www.weforum.org/agenda/2019/04/business-case-for-diversity-in-the-workplace/.

Will Robots Take My Job? https://willrobotstakemyjob.com.

"Yanis Varoufakis Imagines a Post-Capitalist World." ABC Radio National, 15 Oct. 2020, https://www.abc.net.au/radionational/programs/latenightlive/yanis-varoufakis-post-capitalist-world/12769298.

Young, Stephen, and Patrick Kulesa. "Breakthrough Research on Employee Experience." Willis Towers Watson, 2019, https://www.willistowerswatson.-com/en-GB/Insights/campaigns

Made in United States
Orlando, FL
04 November 2026

90200958R00164